REFERENCE LIBRARIES

Education & Libraries

This book is for consultation in the library and may not be borrowed for home reading.

The librarian will be pleased to help with any enquiry or search for information.

Headquarters: Bythesea Road, Trowbridge, BA14 8BS.

690.21

3 836833 000

Tim Freeman, MA (Cantab), CEng, MICE, is Managing Director of Geo-Serv Ltd and former Head of the Foundations Section of the Building Research Establishment.

He has over 16 years of research experience with the geotechnics division of BRE, including 6 years as head of the section studying the performance of foundations for low-rise buildings. He has recently formed his own company specialising in the investigation and remedy of subsidence damage.

Stuart Littlejohn, BSc(Eng), PhD, FEng, FICE, FIStructE, FGS, FRSA, is Professor of Civil Engineering at the University of Bradford and former Chairman of the Ground Board of the Institution of Civil Engineers

He has more than 30 years of industrial experience and is author of many publications covering subjects such as subsidence, structure–soil interaction, site investigation, ground improvement and underpinning.

Richard Driscoll, MSc, FICE, is Head of the Geotechnics Division at the Building Research Establishment.

He has been responsible for all BRE's recent research and publications on the subsidence and heave of buildings on clay soils. He is currently supervising the formation of a national database of subsidence damage cases.

Has your house got cracks?
A guide to subsidence and heave of buildings on clay

T. J. Freeman, G. S. Littlejohn and
R. M. C. Driscoll

Institution of Civil Engineers
and
Building Research Establishment

Published for the Institution of Civil Engineers and the Building Research
Establishment by Thomas Telford Services Ltd, Thomas Telford House,
1 Heron Quay, London E14 4JD

First published 1994
Reprinted 1995, 1999, 2000

A CIP record for this book is available from the British Library

ISBN 07277 1996 3

© Crown and the Institution of Civil Engineers, 1994

Classification
Availability: Unrestricted
Content: Guidance based on best current practice
Status: Established knowledge
User: Home owner

All rights, including translation, reserved. Except for fair copying, no part of this publication may be reproduced, stored in a retrieval system or transmitted in any form or by any means, electronic, mechanical, photo-copying, recording or otherwise, without the prior written permission of the Publisher, Books, Publications Division, Thomas Telford Services Ltd, Thomas Telford House, 1 Heron Quay, London E14 4JD.

The information contained in this book is intended for use as a general statement and guide only. The publishers cannot accept any liability for any loss or damage which may be suffered by any person as a result of the use in any way of the information contained herein.

Typeset in Palatino 10/12 using Ventura 4 at Thomas Telford Services Ltd
Printed in Great Britain by The Cromwell Press, Trowbridge, Wiltshire

Foreword from the Insurance Ombudsman

Recent years have seen a boom in the level of claims for heave and subsidence damage to houses and other low-rise buildings, the vast majority for buildings founded on clay soils. The unusually dry weather of 1989–90 resulted in a sevenfold increase in the level of insurance claims, taking the annual amount spent on the remedy and repair of these defects to more than £500M. Much of the money was spent on *underpinning*, a technique for stabilising the existing foundations.

Yet there is very little published guidance to help the home owner understand why a property founded on clay soil is at risk or to explain how such damage should be remedied, or prevented in the first place.

Consequently, many owners of properties in areas where clay soils are common have become increasingly anxious that their homes may be at risk. Understandably, the smallest of cracks can cause concern; home owners are often confused by conflicting advice on whether their property needs underpinning, and do not know where to go for impartial guidance.

This guide aims to explain in simple terms why properties founded on clay soils suffer cracking, and to give you, the home owner, a comprehensive overview of the subject. It answers the questions which most often worry home owners, such as: how can I be sure the cracks have been caused by subsidence or heave? How bad do they have to be before I need to worry? Should I report them to my insurers? Can I do anything to prevent the damage worsening? Is there a risk of the tree in my garden causing damage? Should I cut it down?

This guide has been sponsored jointly by the Ground Board of the Institution of Civil Engineers and by the Geotechnics Division of the Building Research Establishment. I recommend this publication to home owners as it provides a valuable insight into the nature of building subsidence and heave due to the shrinkage and swelling of clays. Typical preventative measures are also highlighted together with procedures related to insurance claims and remedial work.

Julian Farrand, LLD, Solicitor
Insurance Ombudsman

Preface

This publication has been commissioned by the Institution of Civil Engineers (ICE) and the Building Research Establishment (BRE) in order to provide practical guidance to owners of homes founded on shrinkable clays.

ICE and BRE wish to thank the authors, Tim Freeman (Managing Director of Geo-Serv Ltd and former head of the Foundations Section of BRE), Stuart Littlejohn (Professor of Civil Engineering at the University of Bradford and former Chairman of the ICE Ground Board) and Richard Driscoll (Head of the Geotechnics Division at BRE), who have devoted much time and effort to these deliberations.

The authors are deeply indebted to the many organisations who provided data. They also wish to thank Jan and Keith Sorsby (home owners) and representatives from the insurance industry, who forwarded valuable comments during the final drafting of the guide.

The publishers wish to thank David Fowler for his valuable editorial advice.

Contents

SUBSIDENCE: KEY FACTS	1
1. INTRODUCTION	5
Background	5
Attitudes to subsidence and heave	7
Insurance cover	9
What is covered?	9
Geographical loading	10
2. FACTORS AFFECTING SUBSIDENCE AND HEAVE DAMAGE	12
What are shrinkable clays?	13
Why clay soils shrink and swell	13
Characteristic properties	13
Where are shrinkable clays found?	14
How clay soils behave	16
Shrinkage potential	16
Desiccation	16
How much desiccation	16
Effect of trees	18
Effect of climate	20
Effect of surroundings	23
How your house is built	23
Foundations	24
Floors	27
Walls	29
3. CAN I PREVENT DAMAGE?	32

	Tree management	32
	Structural alterations	32
	Landscaping	33
	Excavations	33
	Drainage	33
4.	**HOW CAN I RECOGNISE DAMAGE?**	35
	Have the foundations moved?	36
	Appearance	38
	Location	40
	Timing	40
	Other indications	40
	Assessing the damage	41
	Cause for concern?	43
	Is stability affected?	44
	Is there a threat to safety?	44
	Is serviceability affected?	44
	Will I be unable to sell the house?	45
	Aesthetics	45
	What should I do?	
	Damage category 0 or 1	46
	Damage category 2	46
	Damage category 3	47
	Damage category 4 or 5	47
5.	**MAKING A CLAIM**	48
	Professional advice	48
	What to expect	52
6.	**THE INVESTIGATION**	55
	Initial inspection	55
	Distortion survey	57
	Desk study	59

	Trial pits	60
	Boreholes	61
	Drain survey	61
	What caused the movement?	62
	Soil type	62
	Foundations	63
	Pattern of movement	63
	Trees	63
	Water	63
	Surroundings	63
7.	**MONITORING**	65
	What is monitoring?	65
	Monitoring to establish cause of damage	65
	Monitoring to measure rate of movement	65
	Monitoring to check success of remedial action	66
	Monitoring techniques	66
	Crack-width monitoring	66
	Monitoring levels	71
	Monitoring lateral movement	71
	Observation period	72
8.	**PREVENTING FURTHER DAMAGE**	73
	Repairing or strengthening the superstructure	74
	Reducing the influence of trees	75
	Tree removal	75
	Tree pruning	77
	Root pruning	78
	Root barriers	79
	Soil stabilisation	79
	Remedial underpinning	80

9.	**IS UNDERPINNING THE RIGHT SOLUTION?**	81
	Criteria for underpinning	81
	Is structural stability threatened?	81
	Is movement continuing?	82
	Is the movement excessive?	82
	What is the cost?	82
	Is underpinning needed?	84
10.	**WHICH TYPE OF UNDERPINNING?**	85
	Mass concrete	85
	Pier-and-beam	87
	Pile-and-beam and piled raft	88
	Mini-piling	91
	Partial underpinning	92
11.	**HAVING THE WORK DONE**	94
	Specification	94
	The Contract	94
	Will I have to move?	96
	Payment	96
	Supervision	97
	Building Regulations	98
	Warranties	98
12.	**WHAT IF THINGS GO WRONG?**	99
	REFERENCES	102
	APPENDICES	104
A	GLOSSARY OF TERMS	104
B	PROFESSIONAL ORGANISATIONS	109
C	FOUNDATION AND SUPERSTRUCTURE DESIGN	111

Subsidence: key facts

This short summary aims to answer the questions which most commonly worry home owners, and to give you a quick overview of the subject. Each answer tells you where in the guide to find further detail, if you need it. Terms shown in *italics* are explained in the Glossary.

Why do properties founded on clay soils suffer subsidence and heave damage?
Most damage is influenced by a combination of four main factors: soil type, weather, vegetation and foundation depth. Many clay soils are classified as *shrinkable*, because they shrink as their *moisture content* decreases and swell as it increases. Shrinkage causes *subsidence* and swelling causes *heave*.

Shrinkage occurs as a result of desiccation, that is the drying caused by evaporation and the extraction of moisture through the roots of vegetation, especially trees. The degree of desiccation in the soil tends to increase during the summer and reduce during the winter. It is greatest following a prolonged period of hot, dry weather and this is often when subsidence damage first appears. Where there are no trees or other large vegetation, desiccation is limited to the top metre or so of the soil and can be reversed by winter rainfall. Trees, however, can extract moisture from greater depths and leave the soil permanently desiccated. The growth of a tree can therefore cause long-term shrinkage and hence subsidence. Conversely, when a tree is removed, the gradual reversal of desiccation can cause long-term heave.

Deeper foundations can be used to protect houses from these ground movements, even where there are large trees nearby. However, most older houses and houses built in areas that were originally devoid of trees are likely to have relatively shallow foundations. For further details see Chapter 2.

How can cracks caused by foundation movement be distinguished from damage due to other causes?

Where damage is relatively slight (for example, no more than Category 2 in Table III, p.39) it is often difficult to identify the cause with certainty. Nevertheless, there are a number of characteristics that can help to distinguish subsidence and heave damage from other causes and these are summarised in Table III, p.39. Where there is any doubt, a *distortion survey* should establish whether or not the walls are leaning from the vertical or if brick courses are no longer horizontal. For further details see Chapter 4.

Can I do anything to reduce the risk of damage occurring or to prevent existing damage from worsening?

For existing properties, care should be taken to keep certain trees, that are known to cause damage, at a sensible size if they are close to the building (see Table I, p.20). Care should also be taken when planting new trees or removing large trees close to the building. Carrying out structural alterations or excavations near foundations can make a property more susceptible to damage and laying drives or paths can alter desiccation by reducing the supply of rain water (see Chapter 2).

For new buildings and extensions, suitable foundation design complying with current regulations and guidelines will dramatically reduce the risk of heave and subsidence damage (see Appendix C).

Are nearby trees causing problems or are they likely to cause damage in the future?

As a rule of thumb, the more damaging trees should be kept at least one tree height away from buildings founded on shrinkable clay; when planting a tree you should therefore take its mature height into consideration. Broad leaf trees are more likely to cause damage than evergreens (see Table I): because of their high moisture demand, oak, elm, willow and poplar are notorious. For further details see Chapter 2.

Should I prune trees, or remove them altogether?

Where a tree is thought to be causing a problem and it is younger than any part of the house, it is normally safe to remove it altogether. Where the tree is older than the house, or any additions to it, do not remove the tree without seeking professional advice.

A GUIDE TO SUBSIDENCE AND HEAVE OF BUILDINGS ON CLAY 3

In such cases, pruning — *crown reduction* or *crown thinning* rather than *pollarding* — is likely to be preferable (see Chapter 8).

When should I start to worry about the damage?
In the vast majority of cases, cracks caused by clay shrinkage during exceptionally dry weather are unlikely to be of structural significance, except for vulnerable features such as brick arch *lintels*. Many cracks will close once there is a return to wetter weather and they can then be repaired as part of routine maintenance and decoration. Where damage is Category 3 or more in severity (see Table III, p.39), or if the problem is not thought to be caused by clay shrinkage, seek urgently professional advice (see Chapter 4).

When should I tell my insurers and how do I make a claim?
You should report to your insurers any damage which is thought to be the result of foundation movement. Table II, p.36, will help you decide if foundation movement is to blame. You can report damage without making a claim and, where cracks are not severe (see previous question), you may simply state that you intend to repair them.

If you do make a claim, your insurer may appoint a loss adjuster whose first task is to establish whether or not the claim is valid. If it is, a suitably qualified professional such as an engineer will be required to investigate the cause of the damage and advise on remedial measures. Many insurers appoint this investigator directly. For further details see Chapter 5.

What will it cost me to make a claim and what is covered?
The cost to you of making a claim is normally limited to the policy excess, generally either £500 or £1000. Where the investigator is appointed by the insurer you will only have to pay this amount if it is decided that there is a valid problem and that repairs are needed. Where liability for damage is denied by your insurers, any costs that you incur will be at your own risk until such time as the claim is proved.

Under the cover provided by most 'Buildings' policies you are entitled to 'reinstatement'. This means that the house will be returned to its pre-damaged condition without any deduction for wear and tear, provided that the house has been well maintained and that the level of insurance is adequate. Insurers will not pay for *underpinning* simply because the existing foundations do not meet current guidelines for building on shrinkable clays. Where it is probable that further damage can be avoided by reducing or removing nearby trees that belong to you, they will normally insist you do this at your expense. For further details see Chapter 1.

Insurers will also wish to satisfy themselves regarding the age of the damage and whether previous problems were disclosed to them when the policy was taken out. A failure to disclose such information or comply with other insurance policy conditions may entitle insurers to reject the claim.

What investigations will be needed and how will it be decided whether or not to underpin?

The investigation generally takes three stages: an initial visual inspection; a more detailed inspection, often involving the excavation of *trial pits* or boreholes and measurements of the distortion in external walls; and a period of *monitoring*. Although monitoring inevitably prolongs the settlement of the claim, it is generally an essential step in deciding how to prevent further damage. It will probably be specified in most cases where the damage is unlikely to threaten safety or the structural stability of the property; occasionally, however, a thorough investigation of soil properties may prove that underpinning is required, thereby removing the need for monitoring. Although there is no generally agreed method for deciding when underpinning is justified, in recent years a number of criteria have been developed (see Table IV, p.83), which are heavily dependent on the results of monitoring. For further details see Chapters 6 to 9.

Chapters 11 and 12 explain how the work will be done and how to complain if something goes wrong.

Chapter 1. Introduction

Much of the UK's housing stock is founded on what are known as *shrinkable clays*. Such clays are strong enough to support a building of four storeys on a simple foundation, but they shrink when their *moisture content* decreases and swell when it increases. Slight movement of houses founded on these soils is therefore inevitable as a result of seasonal changes in moisture content: downward movement or *subsidence* occurring during the summer and upward movement or *heave* during the winter. Normally the whole house is affected and the home owner is unaware that any movement is taking place. Greater movements may occur during long periods of dry weather, leading to sticking doors and windows, or cracking of brick walls.

Only rarely are movements severe enough to threaten the ability of the building to perform its function of safely supporting the loads it has to carry from roofs, floors and wind, without excessive deformation. Movements as severe as this are usually associated with trees (or, occasionally, large shrubs) whose roots extract moisture from the soil. Conversely, removing a tree causes heave as moisture gradually returns to the soil. Large, broad leafed trees are notorious for causing damage.

The susceptibility of a building to cracking as a result of ground movement depends on various factors, the most important of which are the depth of the foundations and the method of construction. Many old houses, such as Victorian and Edwardian ones, were built using lime mortar, which allows bricks to move relative to one another and makes the walls quite flexible. Houses built using lime mortar can often cope with *differential settlements* of 40 mm or more without showing any obvious signs of distress, such as cracking. However, if substantially refurbished, such houses may be as sensitive to movements as modern ones.

Between the First and Second World Wars, cement-based mortars replaced lime mortars. Though cement mortars have many advantages, including strength and durability, they are more brittle so that the walls of modern houses are more easily damaged by foundation movement.

At the same time, as more has been learnt about the risks associated with building on clay soils, there has been a tendency to use deeper foundations to reduce movements. Before the Second World War, it was common practice to use foundations that were no more than 0.45 m deep. Houses built within the past 20 to 25 years, on the other hand, should comply with current guidelines issued by the National House-Building Council (NHBC)[1] and the British Standards Institution (BSI)[2,3], which require a minimum depth of 0.9 m for any buildings founded on clay and deeper foundations where there are trees nearby. These recommendations should ensure that the movements associated with shrinkage and swelling are reduced to levels that can be tolerated by the building.

Because of the link between clay shrinkage and the weather, insurance claims for subsidence damage increase in long dry periods. Figure 1 shows the trend in claims for subsidence and heave damage since the early 1970s when insurance companies first started to cover these risks. There is an obvious peak in the number of claims corresponding to the drought of 1975–76, which brought the terms subsidence and underpinning into common usage. There is also a less well-defined peak in 1984, which was a drier than average year.

Ignoring these peaks, the upward trend in claims between 1972 and 1988 can be seen as a learning curve, reflecting increasing awareness among home owners that they could claim on their insurance for damage due to foundation movement. On the basis of this trend up to 1989 it could be argued that the 'background level' for subsidence and heave damage in the UK costs about £80M to £100M annually. However, as Figure 1 shows, events since 1989 make the earlier peaks pale into insignificance. Between 1989 and 1990 there was a sevenfold increase in the annual value of claims taking the amounts to over £500M in both 1990 and 1991. Even in the wetter 1992, claims reached £260M. The explanation for this boom is that both 1989 and 1990 were unusually hot, dry years, which had a cumulative effect on moisture levels in clay soils possibly without precedent this century; this process is described in more detail in **How clay soils behave, Chapter 2/p.12.**

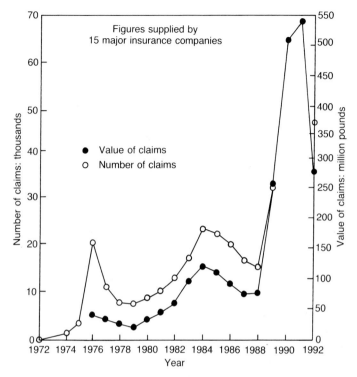

Fig. 1. Trend in claims for heave and subsidence damage to domestic properties. (BRE. Crown Copyright)

Attitudes to subsidence and heave damage

When insurance companies, in response to building societies anxious to safeguard the security for their loans, added subsidence and later heave to the cover included under household policies, it was assumed that claims under this heading would be relatively few — indeed, it is often stated that there was no additional premium for this cover. However, although relatively few buildings are so severely damaged that underpinning is essential for structural reasons, insurance policies do not specify how much movement is sufficient for a valid subsidence or heave claim. Consequently, many home owners have made legitimate claims for relatively minor damage which, before subsidence and heave cover was introduced, would probably have been simply repaired as part of routine redecoration and maintenance. Moreover, in many cases, engineers and surveyors who were concerned that they could not guarantee the damage would not recur erred on the safe side and specified expensive underpinning solutions.

Because of the high cost of underpinning, insurers are reluctant to extend cover to a property that is thought to be at risk, and building surveyors acting on behalf of prospective buyers or mortgage lenders have grown increasingly cautious of recommending the purchase of any property that contains signs of foundation movement. Consequently, anyone trying to sell a house with cracks is faced with the prospect of having to put right what in many cases is no more than cosmetic damage. Where foundation movements are ongoing the only remedy that may be acceptable to buyers and insurers is underpinning, even where movement is relatively small and there is no threat to the structural stability of the building. At the same time, in view of the high level of claims, insurers have hardened their attitudes to underpinning and are now reluctant to sanction underpinning without evidence that the movements are continuing and are likely to lead to a progressive worsening of the damage. *Monitoring* is therefore often specified to obtain the required information, although this inevitably leads to delays in settling the claim, and in the meantime the home owner will find difficulties in selling the property. Although delays are frustrating it is important to settle the claim correctly and monitoring can be an invaluable investigative aid.

There is a groundswell of opinion among insurers, lenders and other professionals connected with the repair of damaged properties that current practice needs revision. Many insurers are now prepared to transfer to a new owner existing cover on a property that has suffered slight cracking, thereby minimising the loss in the value of the house associated with the damage. Recognising that the decision on whether or not to underpin a building should be based on technical grounds, some insurers are also prepared to spend more on an investigation to obtain the necessary information — often removing the burden of these costs from the home owner.

Because cracks are caused by changes taking place in the ground rather than the structure, there is also a growing tendency to involve specialist engineers with a knowledge of soil behaviour (geotechnical engineers) in the decision- making.

In the long term, however, a more sensible approach to repairing and remedying subsidence and heave damage will require a change in attitude among home owners and a return to the values held before insurance policies were introduced to cover the repair of these defects.

Subsidence damage is not new; it has existed for as long as buildings have been built on clay and, although steps are being taken to make new houses less susceptible to foundation movement, it will never disappear completely. For example, trees will continue to be planted too close to properties. Before the introduction of insurance cover in the early 1970s, most home owners regarded minor cracking as harmless and in no way detracting from the value of the house.

Insurance cover
What is covered?

Foundations can move downwards, upwards or sideways; in insurance policies these movements are described respectively as subsidence, heave and *landslip*.

Although the wording in the insurance policy will vary from company to company, unless there is a special endorsement to the contrary, most domestic buildings policies provide cover for damage to the building caused by foundation movement. A typical description of the cover from a buildings policy is:

- damage to buildings caused by subsidence and/or heave of the site on which the buildings stand and/or landslip.

Cover is limited to subsidence and heave 'of the site', to exclude damage caused by movements within the building itself, although this has led to confusion about what exactly constitutes 'the site'. The position was clarified by the Insurance Ombudsman in his 1984 Annual Report, where he defined the site as the prepared ground (including made up ground) on which the building is erected after the foundation trenches have been dug and immediately prior to the first step in the actual process of building. Insurers generally accept this definition, although some policies specifically exclude damage caused by *settlement* and shrinkage of newly made-up ground and one excludes damage caused by initial bedding down of the foundations. Some other common exclusions are:

- loss or damage to garden walls, gates, fences, swimming pools, tennis courts, drives or paving unless buildings are damaged at the same time and by the same cause;
- damage to solid floors caused by compaction of infill or the use of defective materials or faulty workmanship;

- loss or damage caused by river or sea erosion.

It is therefore always important to check the policy wording to see what is and what is not covered. If you do not understand the wording, ask your broker or building society for advice, or write to the insurers for clarification.

Subject to qualifications related to adequacy of sum insured and condition of the building, the cover normally provided in building policies is for **reinstatement**. This covers the cost of physically repairing or replacing the damaged parts of the building, in order to restore it to its original condition, with no deduction for wear and tear.

In practice, it would not be sensible to try to repair the fabric of a building if there is an unacceptable risk of future foundation movements causing further significant damage. Insurers will therefore generally pay for remedial measures such as underpinning, if it can be shown that this is essential to restore the stability of the foundations. However, they will not pay for underpinning simply because the existing foundations are perceived to be inadequate or because they do not comply with current guidelines for building on shrinkable clays. Equally, where it is reasonably probable that further damage can be avoided by reducing or removing nearby trees that belong to you as home owner, insurers normally insist that such work is carried out at your expense, as an item of normal household maintenance. As a general guide it is worth remembering that policies cover repair of damage that has already occurred, not work undertaken purely to prevent future damage.

Insurers will be concerned to establish that all significant damage has occurred whilst they have insured the property and that their position has not been made worse by a failure of a house owner to take reasonable steps or comply with any other conditions noted under the insurance policy.

All policies make you responsible for paying the first part of any claim for subsidence, heave or landslip damage. This sum, known as the excess, had generally been £500 since the risks were first added in the early 1970s, although almost all companies have recently increased the figure to £1000.

Geographical loading

Recent increases in the number of claims for subsidence and heave damage have inevitably led to increases in building insurance premiums. Some insurers have also taken the step of applying geographical loadings i.e. premium variations depending on whether or not the property is located in what is judged to be a high risk area. At present, these loadings are based on postal codes; high risk areas are identified by the number of previous claims and the predominance of shrinkable clay, determined by British Geological Survey maps. Unfortunately, soil type can vary greatly over short distances so that many houses in an area where shrinkable clay is prevalent could, in fact, be founded on comparatively stable sands or gravels.

Another potential inequity of geographical loadings is that they take no account of the foundation design. There is no reason why a foundation cannot be designed to resist movement caused by swelling and shrinkage of clay soils, even close to large trees. The Building Research Establishment (BRE), for example, has for 50 years been advocating the use of *short bored piles* for buildings founded on shrinkable soil; this type of foundation offers greater protection against shrinkage and heave, and need not necessarily be any more expensive than conventional *strip footings* or *trench-fill* foundations (see **How your house is built,** p.23).

If your house is situated in a high risk area comprising shrinkable clays, you may be able to persuade your insurer not to apply a geographical loading because your house is founded on a stable soil such as sand, or is built on piled foundations, and therefore not susceptible to the effects of clay movements. A good insurance broker may be able to help you.

Chapter 2. What causes subsidence and heave?

Foundations are the supporting link between the building and the ground. They transmit the loads from the walls, floors and roof into the ground. At the same time they transfer any ground movement back to the structure, possibly causing distortions and damage. To perform satisfactorily, the foundations must withstand ground movement and limit distortion of the building to tolerable levels. Foundations can fail to do this either because they are inadequately designed for the loads they have to carry or, more commonly, because the ground movements are greater than anticipated. The building then suffers from cracking.

Foundation movement may result from a wide range of factors, which include:

- compression of a soft layer in the ground as a result of the applied foundation loads;
- erosion;
- soil softening;
- variations in the groundwater level;
- compression of filled ground;
- collapse of mine workings or natural cavities;
- nearby construction or excavation;
- frost heave;
- chemical attack on the foundations;
- vibration.

However, by far the commonest cause of foundation movement in the UK is shrinkage or swelling of clay soils caused by changes in the moisture content of the layer of soil near the surface. The remainder of this chapter explains why clay soils change in volume and the circumstances under which damage is most likely to occur.

What are shrinkable clays?

Why clay soils shrink and swell

Clay soils contain a high proportion of extremely small particles with diameters of less than 0.002 mm. Many of these particles consist of one of the three common clay minerals: *kaolinite, illite* and *montmorillonite*. Their molecular structure is such that their crystals, seen under an electron microscope, are shaped like plates. Unlike coarser-grained soils where any water in the ground simply fills the voids between the grains, these small plates can hold the water within their molecular structure, much as a jelly does. An increase in moisture content forces the plates apart causing the soil to expand and, conversely, a reduction in moisture content allows the plates to adopt a denser packing causing the soil to shrink.

The moisture content of a clay soil can be reduced in one of only two ways: by increasing the load on the soil by, for example, constructing a foundation or by raising the ground level; or through moisture being sucked out by evaporation and the roots of vegetation — a process known as *desiccation*.

If the applied load on the soil is reduced, or the source of suction is removed, moisture will be drawn back into the soil. Whether the clay is swelling or shrinking, because of its limited *permeability* these volume changes occur only slowly, over a period of months or even years.

Characteristic properties

Clays are characteristically mouldable (or plastic) and smooth and greasy to the touch. The more clay particles in the soil, compared with *silt* or other coarser-grained material, then the more pronounced these characteristics are. Each clay mineral has its own characteristics; montmorillonite, for example, can absorb far more moisture than either illite or kaolinite.

The strength of clays in their natural state can vary from 'soft', through 'firm' and 'stiff' to 'hard'. These classifications have precise definitions,[4] but as a rough guide, you can easily mould a sample of soft clay in your hand, while it is only just possible to push your thumb nail into a sample of hard clay. The variation in strength of near-surface clays is largely due to their geological history. The stronger clays tend to be the older ones which had hundreds of metres of material subsequently deposited over them; this cover was later removed by erosion and glacial action in the Ice Age. Such clays are described as *overconsolidated* and, near the ground surface, would normally be classified as 'firm'.

Although mouldable when wet, firm clays shrink and crack as they dry and intact lumps can become very hard to break. A clod of firm clay immersed in water will soften only slowly, without disintegrating. A clod which disintegrates quickly contains silt and other coarser-grained materials.

Liquid and plastic limits

A given clay soil can be characterised by measuring the range of moisture contents over which it is mouldable. The upper end of this range where the soil begins to 'flow' is known as the *liquid limit* and the lower end where the soil begins to break up is known as the *plastic limit*. As with all soil moisture contents, both limits are expressed in gravimetric terms, that is as the weight of water removed by drying as a percentage of the weight of dry material.

Standard laboratory tests have been defined to measure these limits on a consistent basis[5] so that they can be used to compare different clays in terms of their likely behaviour. The liquid limit, for example, is a measure of the amount of water bound to soil particles; hence, the greater it is, the more 'clayey' the behaviour of the soil.

Where are shrinkable clays found?

Firm shrinkable clays, capable of supporting buildings of up to three or four storeys on shallow foundations, occur widely in the south east of England, as shown on the map in Figure 2. The geological names of these clays include: London; Gault; Weald; Kimmeridge; Oxford; Woolwich & Reading; Lias; Barton; and the glacial drift clays, such as the chalky boulder clays of East Anglia, in which clay has been mixed with a range of other soils during the Ice Age. Their moisture contents are close to the plastic limit; for example, a typical moisture content for weathered London Clay with a plastic limit of say 26% would be in the range of 25 to 30%. However, close to the ground surface, the moisture contents are influenced by evaporation and rainfall, and fluctuations from as

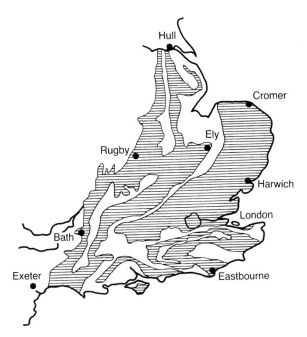

Fig. 2. Distribution of shrinkable clays in South-East England

little as 15% in dry summer weather to 40% in wet winters can occur.

Some shrinkable clays occur further north than the areas indicated in Figure 2; for example, those derived from the weathering and glaciation of Carboniferous shales around Sunderland and north of Shrewsbury. However, in the North the surface clays are generally sandy and their potential for shrinkage is, therefore, smaller. In addition to the firm clays, there are soft, alluvial clays found in and around estuaries, lakes and river courses, such as the Fens, the Somerset levels, the Kent and Essex marshes, and the Firths of Forth and Clyde. All these clays have a firm, shrunken crust which is drier than the body of clay beneath. Clay shrinkage is not the only foundation problem in these areas: excessive settlement due to loading the underlying softer clay and *peat* can also occur. More detailed information on the location and identification of clay soils can be obtained from British Geological Survey maps.[6]

How clay soils behave

Shrinkage potential

The potential for a clay soil to cause damage by shrinking or swelling is called its *shrinkage potential*. This parameter is usually assumed to be proportional to the difference between the liquid and plastic limits, a quantity which is known as the *plasticity index* or simply the *plasticity* of the clay.

Three classifications of shrinkage potential (low to high) are suggested by the National House-Building Council (NHBC),[1] but BRE have more recently proposed the following:

Classification	Plasticity index: %
low	0 to 20
medium	20 to 40
high	40 to 60
very high	over 60

A more fundamental approach to understanding, and hence calculating, the volume changes and movements associated with clay shrinkage and swelling depends on a more detailed examination of the process of desiccation.

Desiccation

Most people associate the term desiccation with a complete removal of moisture; a 'desiccator' for example is a device for removing the last traces of free water from crystals and 'desiccated' coconut is an example of a food that has been thoroughly dried in order to preserve it. To geotechnical specialists, however, the term has a slightly different meaning: it is **any** reduction below the soil's natural moisture content caused by evaporation or removal of water through tree roots. A soil described as desiccated may in fact have a moisture content that is reduced by only 1% or 2%.

Desiccation often reduces the moisture content of firm shrinkable clays to values close to or below the plastic limit. This makes the soil appear dry and hard and, near the surface, often causes the ground to crack. In highly shrinkable soils, cracks of 25 mm wide and 0.75 m deep are not uncommon during dry summers.

How much desiccation?

Deciding how much, if any, desiccation has occurred is not easy. In highly uniform ground it should be possible to see any differences

A GUIDE TO SUBSIDENCE AND HEAVE OF BUILDINGS ON CLAY   17

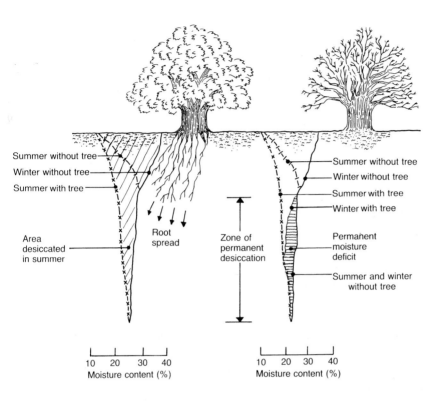

Fig. 3. Seasonal variation in moisture content with and without trees

attributable to desiccation, provided soil moisture content measurements are available both at the suspect location and remote from it; unfortunately, clay soils are rarely sufficiently uniform for this procedure to be reliable. Furthermore, the luxury of information from both the site and remote from it is unusual. More often investigators have had to make-do with moisture content values from the suspect location only; this approach is fraught with uncertainty and has often led to very cautious conclusions about the cause and extent of subsidence.

However, the BRE has recently developed the 'filter-paper test', a relatively simple laboratory procedure[7] to measure the state of *suction* and desiccation in clay samples. The results of this test are usually a

Fig. 4. Dramatic example of the potential effects of tree growth on a house with shallow foundations. (BRE. Crown copyright)

much more reliable guide to the amount of desiccation and should provide better assessments of the need for, and extent of, any underpinning, and the need for any action against nearby trees.

Effect of trees

Work at the BRE has shown that, in grass-covered areas, the effect of evaporation in firm, shrinkable clays is largely confined to the uppermost 1 m to 1.5 m of soil. However, where there are trees, and to a lesser extent hedges and large shrubs, moisture can be extracted from depths of 6 m or more. For high-plasticity clays which tend to have very low permeabilities, rainfall during winter cannot fully replenish the moisture removed by large trees during the summer. Hence a zone of permanently desiccated soil develops under the tree, as shown in Figure 3.

As the tree grows, the desiccated zone increases in depth and width, producing more subsidence which is likely to affect any nearby structures. In some instances, the subsidence associated with a growing tree can be dramatic; Figure 4, for example, shows a shallow founded Victorian property that has cracked as the result of a poplar tree planted at about the time that the house was built. The photograph, taken 50 years later, shows classic symptoms of subsidence increasing towards the tree.

The extent of the desiccated soil depends on the moisture demand of the tree. In general, broad leaf trees have a greater moisture demand than evergreens. Because of their size, oak, elm, willow and poplar are notorious for causing damage. However, these are not the ones which most commonly cause damage to housing, because other trees with lower moisture demands, notably plane, lime and ash, are more frequently planted close to buildings. Information collected by the Royal Botanic Gardens during the 1970s[8] suggests that the trees most likely to cause damage, in descending order of threat, are as listed in Table I.

For each type of tree, the table gives the distance between the tree and the building within which 75% of the reported cases of damage occurred. As a rule of thumb, it would appear that damage can usually be avoided by ensuring that the tree is no closer to the foundations than its mature height. For the less 'thirsty' trees, this figure can be reduced to half the mature height. There are two reservations about this generalisation. First, it takes no account of the shrinkage potential of the soil or the depth of the foundations. Second, it is the leaf area of the tree rather than its height that ultimately determines its moisture demand. The rules should therefore be treated with caution.

Table I. Risk of damage by different varieties of tree [8]

Ranking	Species	Maximum height of tree (H): metres	Separation between tree and building for 75% of cases: metres	Minimum recommended separation in shrinkable clay: metres
1	Oak	16–23	13	1H
2	Poplar	24	15	1H
3	Lime	16–24	8	0.5H
4	Common Ash	23	10	0.5H
5	Plane	25–30	7.5	0.5H
6	Willow	15	11	1H
7	Elm	20–25	12	0.5H
8	Hawthorn	10	7	0.5H
9	Maple/Sycamore	17–24	9	0.5H
10	Cherry/Plum	8	6	1H
11	Beech	20	9	0.5H
12	Birch	12–14	7	0.5H
13	White Beam/Rowan	8–12	7	1H
14	Cypress	18–25	3.5	0.5H

Effect of climate

The degree of desiccation in the soil is greatest towards the end of summer and least in late winter or early spring, and this is reflected in ground movement. To illustrate this point, Figure 5 shows ground movements measured at various depths at a London Clay site over a three year period; results are shown both for a grass-covered area and for an area containing some large poplar trees. The movements were substantially greater in the dry summers of 1989 and 1990 than they were in 1988, confirming that desiccation increases in hot dry weather.

The site measurements also confirm that ground movements in the grass-covered area are generally confined to the surface metre of soil, although the unusually dry weather of 1989 and 1990 did produce movements of 6 mm and 13 mm respectively at a depth of 1 m.

Understandably, the movements in the vicinity of the poplar trees were larger and, for example, at a depth of 1 m, exceeded 35 mm in an average year such as 1988. During 1989 and 1990 measurable

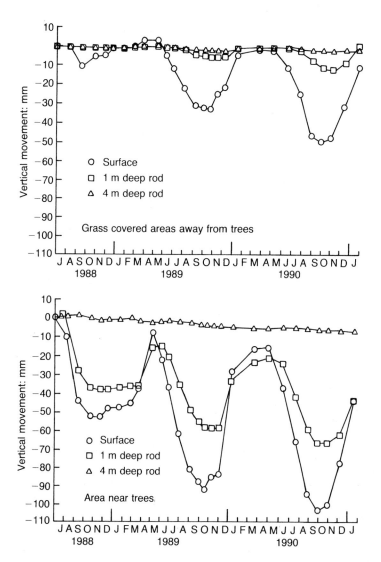

Fig. 5. Measurements of ground movement at various depths in London Clay. (BRE. Crown Copyright)

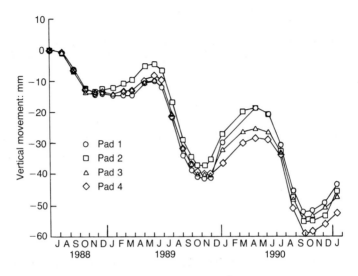

Fig. 6. Observed movement of pads with various applied loadings. (BRE. Crown Copyright)

ground movements were recorded even at depths of 4 m indicating that prolonged periods of dry weather can have an effect on deep-seated desiccation. The potential effect of these ground movements on buildings sited near trees is graphically illustrated by the measurements shown in Figure 6, which were made on some 1.5 m deep concrete pads, as might be used in a factory foundation, for example. These pads were positioned an average of about 5 m from large poplars. The cumulative moisture losses during the dry summers produced a 'ratcheting' effect on the pads, resulting in settlements of more than 50 mm over a three year period. This movement would be in addition to any long-term subsidence associated with the growth of the trees. It would be capable of causing cracks in most brick-built structures, particularly if only part of the building was affected, maximising differential movement and hence distortion.

A GUIDE TO SUBSIDENCE AND HEAVE OF BUILDINGS ON CLAY   23

Effect of surroundings

Desiccation depends on the availability of water, which in turn will depend on a variety of factors. As a home owner you can, to an extent, control some of these factors, such as drainage and the permeability of the surface layer — more details are given in Chapter 3. Other factors, such as the slope of the ground and the shelter provided by the house and other nearby buildings, cannot be controlled.

Ground that slopes steeply away from the house may increase desiccation in the soil under the foundations by lowering the ground water table and making it easier for rain to run off rather than be absorbed by the soil. Although there are no detailed observations to confirm this, it is possible that the consecutive hot summers of 1989 and 1990 caused cumulative movements, similar to those described in *Effect of climate* (p.20), near cuttings, excavations and steep slopes.

How your house is built

It is not uncommon for only one or two houses in a street to be damaged by subsidence, although all are founded on the same type of soil and are similarly close to large trees. Often the explanation lies in local variations in the soil or minor differences in the method of construction, which can have an important influence on the house's susceptibility to subsidence and heave damage.

Construction practice has changed significantly during the course of the 20th century. For example, cavity walls have been introduced to improve insulation, lightweight concrete blocks have replaced bricks for all but the outside leaf of external walls, to reduce material costs and to improve thermal insulation and, with the advent of ready mixed concrete and increased labour costs, mass concrete has largely replaced the use of brickwork below ground level. At the same time, regulations and guidance concerning house construction have kept pace with advances in knowledge and increased demand for economy in relation to construction, maintenance and energy efficiency. These changes have all had an effect on the susceptibility of houses to subsidence and heave damage.

Whereas a comprehensive account of house construction is outside the scope of this guide, the following sections describe some

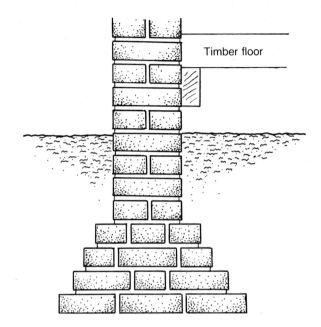

Fig. 7. Cross-section through a typical Victorian house foundation

of the more important differences between older houses and those being built today.

Foundations

In the UK, the concept of providing foundations as a matter of routine for low-rise buildings dates back to the early 19th century. Prior to this it was common practice to build walls either directly onto the bottom of a narrow trench or onto a thin layer of rubble compacted in the bottom of a trench. The Victorians realised that the stability of the building could be improved by spreading the load in the wall over a greater area; they achieved this by stepping or *corbelling* the brickwork at the base of the wall to form a *footing* as shown in Figure 7. In speculative housing, the depth to the underside of the footing from ground level was typically 450 mm unless poor ground forced the builder to dig deeper; in some cases it might be even shallower. By 1890 it was becoming common in better construction to line the bottom of the trench with a layer of weak unreinforced concrete about 600 mm wide and 215 mm deep.

This type of foundation, which is known as a strip footing, is still widely used today, although the practice of using corbelled brickwork has been dispensed with and the stronger, modern-day concrete is relied on to distribute the load. Figure 8 illustrates a typical detail for a modern strip footing showing its use with two different types of floor.

By the late 1940s it was realised that buildings founded at a depth of 450 mm were susceptible to movement as a result of moisture changes in the soil. Consequently, a minimum foundation depth of 0.9 m (3 feet) was proposed for buildings on clay soils. At the same time, it was also realised that it would be cheaper to construct a foundation to this depth by filling a trench, say, 400 mm wide with mass concrete, rather than having to excavate a wider trench that would be needed for a bricklayer to work in. Hence, a new type of foundation, the *trench-fill*, was developed which largely replaced the strip footing for building on clay soils. A typical detail of a modern trench-fill foundation is shown in Figure 9; the depth can easily be increased and depths of 3.5 m or 4 m are sometimes used when building near trees.

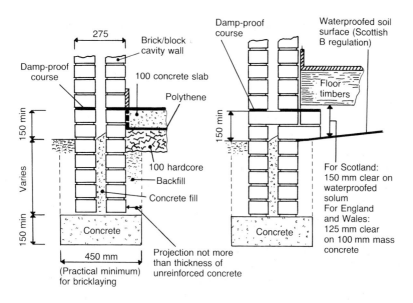

Fig. 8. Typical strip footing shown with ground bearing floor slab (left) and a suspended timber floor (right)

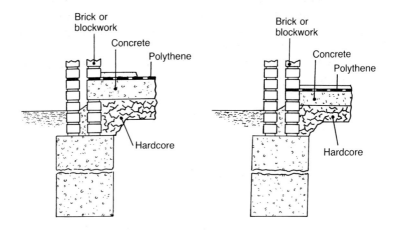

Fig. 9. Trench-fill foundations with two alternative floor slab arrangements

Since it first appeared in a code of practice in 1949,[9] the recommended minimum foundation depth of 0.9 m for building on clay soils has remained unaltered, although the guidance issued by the NHBC,[10] recommended a slightly greater depth of 1.0 m on highly shrinkable clay soils. However, the recommended minimum depth was not generally adopted until the introduction of Building Regulations in 1965 made it mandatory in many areas to comply with the relevant codes. Consequently, many houses built before 1965 may have foundations that are less than 0.9 m deep, despite the obvious presence of shrinkable clay.

Although the 1949 code of practice recognised that deeper foundations were needed near trees, the first document to give specific advice on how the required depth should be determined was the NHBC practice note *Building near trees*,[10] which first appeared in 1969. This note gave a method for calculating the required foundation depth, based on the type and height of the tree and its distance from the proposed building. This was revised in 1974 to take into consideration the shrinkage potential of the soil and the geographical location of the site. Recommended depths have remained unchanged since then and are in the range 0.9 m to 3.5 m.

In summary, over the past 100 years foundations have become progressively deeper, stronger and stiffer, especially for houses near trees. This has both advantages and disadvantages. The advantages are that, because they are deeper, the foundation movements caused by variations in desiccation are smaller and, because they are stronger, the foundations tend to spread the effects of the ground movements over a larger area, reducing distortion. The disadvantage is that, when excessive subsidence or heave does occur, the foundations will tend to crack at one point, concentrating the movement over a short length of wall and producing more severe damage in the superstructure.

Alternative forms of foundation

Although trench-fill construction is possible to depths of 3.5 m or more, it becomes increasingly expensive because of the large amounts of excavation and concrete that are required. Consequently, where foundation depths of several metres are required, it is often more economical to construct isolated *piers* or to install *piles* rather than excavating a continuous deep trench. The tops of

Fig. 10. Beam and block floor (BRE. Crown Copyright)

the piles or piers can then be connected using a reinforced concrete beam (a *ground beam*), which supports the walls.

Another form of foundation used for low-rise buildings is the *raft*, a reinforced concrete slab covering the entire area of the building. The raft, which is often constructed on a bed of compacted *hardcore*, distributes foundation loads and can help reduce distortion in the brickwork as a result of differential ground movement. Rafts are therefore commonly used on soft soils and fill, or in areas prone to mining subsidence. However, their application to construction on shrinkable clays is limited. In areas where there are no large trees and only a minimum foundation depth is needed, a raft is more expensive than conventional construction using trench-fill and a separate floor slab. Near trees, it is necessary to remove the severely desiccated soil and replace it with fill, which involves not only large costs for excavation and removal of spoil, but also careful control of the filling to ensure that it is adequately compacted.

Floors

Floor construction has also changed during the 20th century. In Victorian times the floor of the bottom storey was usually built of timber supported between the foundations and intermediate *sleeper walls*. This *suspended floor* provided an underfloor space, which was sometimes deepened under part of the house to provide a coal cellar, for example. With the advent of concrete as a building material, the suspended timber ground floor was largely superseded by the concrete floor slab. The cheapest way of constructing such a slab is to cast it directly on a layer of compacted hardcore resting on the ground, after removing the topsoil. However, this leaves the floor very sensitive to swelling and shrinkage in the surface soil and is therefore undesirable on shrinkable clay sites, especially where trees or other vegetation have been recently cleared. It is therefore becoming increasingly common to use a suspended floor slab of reinforced concrete, which can span between the foundations without the need for support from the ground. This type of floor is normally constructed either using precast concrete beams with the gaps filled by lightweight concrete blocks (known as a *beam-and-block floor* — see Figure 10), or by casting a slab on a proprietary *void former*.

Walls

Most UK housing is constructed using load-bearing masonry. The commonest form of masonry is brick, although natural stone is popular in some parts of the country and, over the past 30 years, the use of lightweight concrete blocks has become increasingly common, especially for internal walls. The main alternative to load-bearing masonry is a framed construction, in which a load-bearing frame carries the roof and floor loads and some type of

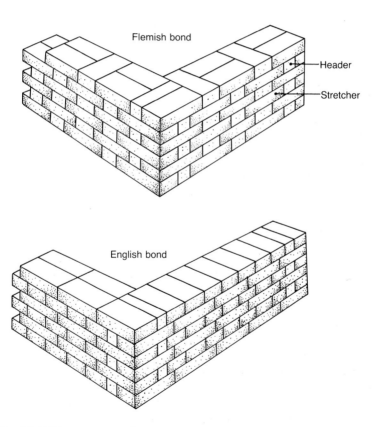

Fig. 11. **Different types of bond for constructing a one brick thick solid wall**

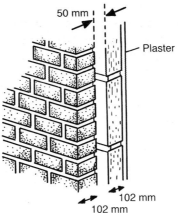

Fig. 12. Cavity wall

external cladding or panel forms the walls. Timber frame houses began to be built in large numbers in the UK during the 1980s though their market share has always been limited. Steel or reinforced concrete frames with infill panels have mainly been used for high rise blocks. However, in the ten years after the Second World War, when there was a shortage of labour and materials and the house building industry was unable to meet demand using traditional methods, some half a million public sector houses were constructed using non-traditional methods. These methods included: concrete posts and infill panels, thin concrete slabs supported on light structural steel frames, pre-assembled brickwork panels, stressed-skin resin-bonded plywood panels, various forms of asbestos sheeting, and curtain walling. As a general rule, a framed building is likely to be more flexible than one constructed from load-bearing masonry **(see Appendix C Foundation and superstructure design .).**

In traditional brick construction, external walls not exceeding two storeys in height were generally built one brick-length (215 mm) thick. For taller walls, a thickness of one and a half or two brick-lengths would be used for the lower stories. To tie the wall together, some bricks (know as *headers*) were laid end on to the outside face of the wall. Bricks with their long side parallel to the face of the wall are called *stretchers*. Different arrangements of headers and stretchers, known as *bonds*, are possible; the two most common in house construction, English and Flemish bond, are illustrated in Figure 11. The bricks were held together with a *lime mortar*, consisting typically of one part lime to three parts sand, which developed its strength slowly and was not a particularly durable material.

The most important change in low-rise building practice during the 20th century has been the introduction of cavity wall construction, where external walls are built using a separate outer and inner 'leaf' divided by an air gap. This form of construction, which significantly reduces damp penetration and improves thermal insulation, was adopted in certain parts of the UK during the 19th century but was not generally accepted as the standard form of load bearing masonry construction until after the Second World War. A cross-section through a typical cavity wall consists of a 102 mm thick brick outer leaf, a 50 mm wide cavity and an inner leaf of 102 mm thick insulating, load-bearing concrete blocks (see Figure 12). The two leaves are tied together by wall ties, which were originally made of wrought iron but are now more likely to be heavily galvanised mild steel, stainless steel or plastic.

The individual leaves of a cavity wall are inherently weaker than a 215 mm solid wall, making it essential that there is a strong bond between the individual bricks and blocks. Consequently, stronger cement-based mortars have largely replaced the softer lime mortar. Ironically, cement mortars make the wall less flexible and more prone to cracking as a result of distortion.

Cavity walls can normally be recognised because they are built entirely of stretchers. Where the outside walls have been rendered or pebble dashed, it will be necessary to measure the thickness; excluding internal or external finishes, cavity walls are generally around 255 mm thick, compared to about 215 mm for a solid wall.

Chapter 3. Can I prevent damage?

Heave and subsidence damage affects only a small proportion of houses in this country and it would be wrong to suggest that, as a home owner, you need to worry unduly about protecting your house from these defects. At the same time, many houses that do suffer cracks are old properties whose foundations have performed satisfactorily for many decades or even centuries. Only occasionally is the damage caused by a problem that has dogged the property periodically since it was first built; more often it is the result of either a lack of necessary maintenance or an injudicious 'improvement'.

The following section describes some of the factors that affect the susceptibility of a house to heave or subsidence damage and which, to a degree, you can control.

Tree management

Because of their effect on moisture content, you should avoid planting trees, hedges or large shrubs close to houses founded on clay soils. Remember to take the fully mature height of any tree you plant into account (see **Effects of trees**, Chapter 2). Further advice on safe planting distances can be found in BS5837 (1992)[3] and in the *Gardening Which* article 'Trees near the house'.[11]

Where existing trees are too close to foundations, you should think about having them pruned or removing them altogether. But removing a tree which is older than the house or any later extensions can be dangerous because it may result in subsequent heave. For more details see **Reducing the influence of trees**, p.76.

Structural alterations

Structural alterations, such as the removal of load-bearing walls or the addition of a second storey to a single storey extension, can alter foundation loads. Whereas this may cause settlement under the more-heavily loaded foundations, the movements on firm clays will be unlikely to cause cracking as they are very small compared to the amount the clay can shrink and swell. The alterations may, however, make the structure more susceptible to damage. For example, modern cavity walls are likely to be less flexible than solid walls built with lime mortar, and converting ground floors to open plan may tend to concentrate distortions at isolated points.

It follows that some alterations may cause cracking in an old property which has never previously suffered heave or subsidence damage. Such damage would be most likely to show up after a period of unusually dry weather, when the differential settlements caused by seasonal variations are greatest. Some insurance policies exclude subsidence damage which arises as a result of structural alterations.

Landscaping

You should take care in planning patios, car parking space, and any other operations that alter ground levels on shrinkable clay sites. These can reduce the effective depth of the foundations, thereby increasing seasonal movements.

As mentioned in **Effects of surroundings** in Chapter 2/p.23, altering the slope of the ground may also affect desiccation.

Excavation

Take care whenever a large pit or trench is dug close to a building, because there is a tendency for the soil to move towards the excavation. In shrinkable soils there is the additional risk that the excavation will reduce the groundwater level locally, thereby increasing desiccation.

Drainage

Water leaking into the ground near foundations as a result of damaged drains is generally undesirable as it can erode the soil and backfill materials; in shrinkable clays it is also likely to have a local effect on desiccation. Near trees, it may reduce desiccation in the short term, although the tree will rapidly grow new roots to exploit the moisture supply, causing further damage to the drain and ultimately blocking it. Where there are no trees, the reduction in desiccation associated with a leaking drain may be detrimental as it could increase differential settlement.

Improving surface or underground drainage tends to lower the water table, which may increase desiccation. Conversely neglecting existing land drains will allow water levels to rise, reducing desiccation. In either case, the changes can cause ground movement which may have a detrimental effect on any nearby structures.

Impermeable coverings such as concrete or asphalt can also affect desiccation, both by reducing evaporation and also by restricting the rate of rainfall infiltration into the soil. Where there are large trees, the net effect may be to force the tree to take moisture from further afield, which will in turn generate ground movements in the surrounding soil; if there is a risk of damage, it may be preferable to use porous pressed concrete slabs which should have less effect on desiccation.

Chapter 4. How can I recognise subsidence damage?

Most home owners do not make a habit of regularly inspecting their houses for cracks; you may therefore not notice minor damage for some time. Often cracks only come to light during decoration or building work or, very commonly, when your house is being surveyed on behalf of a prospective purchaser. Alternatively, you may notice damage when it starts to affect serviceability: for example, windows stick and doors fail to close properly. Whatever the circumstances, having discovered cracks, you are likely to be concerned that your house is suffering heave or subsidence damage, and about the implications. These could range from the damage making it more difficult to sell your house to there being a risk of some of it falling down.

If you find yourself in this position, you have two broad courses of action. The first is to fill the cracks during routine maintenance and consider whether any of the measures described in Chapter 3 can be easily implemented to reduce the likelihood of the damage returning; the second is to seek expert advice either independently or through insurers.

If the damage is severe the choice will be obvious. In the vast majority of cases, however, the damage is likely to be relatively minor and your decision should be based on the following four steps.

- Confirm that the damage is caused by foundation movement.
- Assess the damage objectively.
- Consider the potential consequences of not putting right the cause of the damage
- Decide, depending on how badly your house is affected, whether or not the cost of remedial work is justified.

A more detailed description of these four steps follows.

Have the foundations moved?

Foundation movement is only one of many processes that can cause distortion and cracking in buildings. Other common causes of cracking are frost attack, thermal expansion and contraction, drying shrinkage, over-stressing of walls or floors (e.g. as a result of injudicious structural alterations) and chemical attack. Distinguishing damage due to foundation movement from that caused by these other processes can sometimes be difficult, particularly where the damage is relatively minor. Nevertheless, there are a number of general indicators which can be used to help identify damage caused by foundation movement; these are discussed below and summarised in Table II.

Table II. Indicators of foundation movement

Foundation movement in general
Few isolated cracks at weak points in structure
Cracks taper from top to bottom
Cracking is continuous through dpc
Cracks exceed 3 mm wide
Cracking occurs both externally and internally at the same location
Cracking consistent with a pattern of movement
Doors and windows stick
Wallpaper rucks at corners and between walls and ceiling
Gaps appear below skirting board or between floor boards and wall
Roof tiles displaced, or other signs of distortion in roof
Drains and services disrupted
Walls measurably out of level or out of plumb
Movement due to shrinkable clay
Cracks first appear after prolonged period of dry weather
Cracks open in summer and close in winter
Largest cracks in south-facing walls or near trees
Obvious damage to garden walls and other structures on shallow foundations
Cracking to paving and asphalt around trees

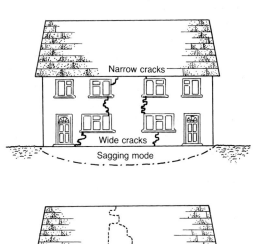
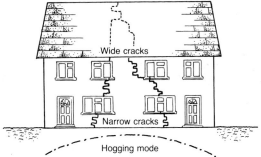
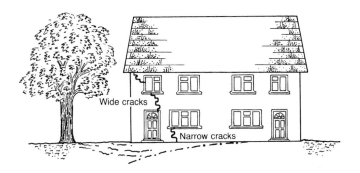

Fig. 13. Different patterns of movement and cracking caused by subsidence and heave

Fig. 14. Classic example of damage (category 3, in this case) caused by hogging distortion. (BRE. Crown Copyright)

Appearance

Foundation movement tends to produce a few, relatively large, isolated cracks, rather than a profusion of small, widely distributed ones. These cracks often taper from top to bottom, reflecting the fact that one part of the structure is rotating relative to another. Considered as a whole, the pattern and taper of the cracking should be consistent with a particular pattern of movement, as shown in Figure 13. Where clay shrinkage is the cause, the pattern of cracking will often indicate increasing subsidence towards a tree, or increasing heave towards the position of a removed tree. Some classic examples of building damage caused by foundation movement are shown in Figures 14 and 15; the categories of damage refer to the BRE classification summarised in Table III.

A GUIDE TO SUBSIDENCE AND HEAVE OF BUILDINGS ON CLAY 39

Category of damage	Description of typical damage (Nature of repair in italic type)
0	Hairline cracking which is normally indistinguishable from other causes such as shrinkage and thermal movement. Typical crack widths 0.1 mm. *No action required.*
1	Fine cracks which *can easily be treated using normal decoration.* Damage generally restricted to internal wall finishes; cracks rarely visible in external brickwork. Typical crack widths up to 1 mm.
2	*Cracks easily filled. Recurrent cracks can be masked by suitable linings.* Cracks not necessarily visible externally; *some external repointing may be required to ensure weather-tightness.* Doors and windows may stick slightly and *require easing and adjusting.* Typical crack widths up to 5 mm.
3	Cracks which *require some opening up and can be patched by a mason. Repointing of external brickwork and possibly a small amount of brickwork to be replaced.* Doors and windows sticking. Service pipes may fracture. Weather-tightness often impaired. Typical crack widths are 5 to 15 mm, or several of, say, 3 mm.
4	Extensive damage which *requires breaking-out and replacing sections of walls,* especially over doors and windows. Windows and door frames distorted, floor sloping noticeably*. Walls leaning or bulging noticeably; some loss of bearing in beams. Service pipes disrupted. Typical cracks widths are 15 to 25 mm, but also depends on number of cracks.
5	Structural damage which *requires a major repair job, involving partial or complete rebuilding.* Beams lose bearing, walls lean badly and require shoring. Windows broken with distortion. Danger of instability. Typical crack widths are greater than 25 mm, but depends on number of cracks.

Table III - BRE - Classification of damage (based on nature of repair). BRE. Crown Copyright

Important Note Crack width is one factor in assessing category of damage and should not be used on its own as a direct measure of it.

*Local deviation of slope, from the horizontal or vertical, of more than 1/100 will normally be clearly visible. Overall deviations in excess of 1/150 are undesirable.

Except very locally, causes of cracking other than foundation movement tend to produce only relatively small cracks, less than 3 mm or so wide. Therefore, although a crack width of more than 3 mm is not a necessary condition for the damage to have been caused by foundation movement, it does help rule out many of the other causes. A pound coin is exactly 3 mm thick and provides a simple way of measuring whether the crack exceeds this width. To be significant the

Fig. 15. Classic example of damage (category 4, in this case) caused by sagging distortion. (BRE. Crown Copyright)

crack should be in a brick or block wall, rather than appearing solely in plasterwork or at the edge of a *stud partition*, for example.

Location

The location of the cracks is as important as their physical appearance. Foundation movement often results in cracks at weak points, such as window openings and doors, or at points where there is a change in foundation depth, such as the junction of a bay or an extension with the main structure. Moreover, whereas many other causes of damage will tend to affect brickwork either only above the *damp-proof course* (dpc) or only below it, foundation movement can produce cracks that are continuous through the dpc. The cracks are often visible from both sides of the wall and foundation movement is one of the few processes that can cause cracking in both leaves of a cavity wall at approximately the same location.

Timing

Damage due to clay shrinkage normally shows up after a long period of dry weather. Subsequently, the cracks will tend to close

A GUIDE TO SUBSIDENCE AND HEAVE OF BUILDINGS ON CLAY 41

in winter, or wet periods, and may re-open during another dry summer.

The presence of shrinkable clay can often be confirmed by the effect of excessive surface movements on garden walls and other structures on shallow foundations. Paving may dip noticeably towards large trees and asphalt surrounding trees may contain crescent shaped cracks as shown in Figure 16.

Other indications

Foundation movement tends to distort openings and often causes doors and windows to stick. In some cases the distortion may also affect partitions, ceilings, floors and the roof, resulting, for example, in rucking of wallpaper in corners and at the junction of walls and ceilings, gaps below skirting boards or slippage between

Fig. 16. Classic example of cracking in asphalt associated with a large tree. (BRE. Crown Copyright)

Fig. 17. Classic example of roof distortion caused by foundation movement. (BRE. Crown Copyright)

roof tiles as shown in Figure 17. Large movements may disrupt services, particularly drains.

The best way of **confirming** that the foundations have moved is to measure how much external walls are out of plumb or how much brick courses are out of level. Brick walls are unlikely to have cracked unless there have been several centimetres of differential settlement, which should produce distortions that can be readily distinguished from any variations in level or plumb due to construction inaccuracies. Further details of the techniques used to measure the amount of movement that has occurred can be found in **Distortion surveys**, Chapter 6.

Assessing the damage

Damage assessment can be very subjective. Words like 'slight', 'bad', 'severe' or 'unacceptable' are often used, but what may seem severe to you as home owner may in fact be slight in terms of its effect on the serviceability or stability of the building. To help reduce this subjectivity and help minimise misunderstandings, a classification of damage has been published in *BRE Digest* 251,[12] which ranges from 0 for slight, cosmetic damage to 5 for damage that is likely to require partial or total rebuilding. This classification, reproduced here in updated form as Table III (p.39), is based on the ease of repairing the damage: it gives no indication of how and when action is needed to improve the stability of the foundations. For example, the table indicates that damage up to Category 2, which may include cracks up to 5 mm wide, can be easily filled and covered by redecoration. Nevertheless, extensive redecoration is itself an expensive operation and might be considered inappropriate if there is a probability of the damage returning within a short time. Equally, a 5 mm crack which appears suddenly may indicate progressive movement that is going to cause further damage unless action is taken to remove the cause.

Cause for concern?

Before you can decide if the cost of work to repair and prevent damage recurring is justified, you need to examine the potential consequences of various levels of damage; these are given below, the most severe first.

Is stability threatened?

Most UK houses have load-bearing masonry walls which carry the roof and floor loads. These loads cause mainly vertical compressive stresses, and the walls' ability to carry such loads remains satisfactory even if they are cracked. Differential foundation movement (variation in the magnitude of movement across the foundation), however, tends to increase the tensile and shear stresses in the wall, and very large movements can make the wall unstable. This is unlikely to occur unless there is **Category 5** damage, but it would effectively make the building uninhabitable. Temporary supports such as shoring would be needed to prevent collapse.

Is there a threat to safety?

It is very rare for clay shrinkage or swelling to cause overall structural instability, but there is a possibility of *lintels* and small sections of brickwork becoming unstable once the damage exceeds **Category 3**. Brick-arch lintels used extensively in Victorian architecture are especially vulnerable, and these can be affected by a crack of no more than 3 mm. Moreover, where the foundation movement causes large rotations, there is a risk of roof and floor *joists* losing their bearing, especially in older houses where joists were often simply built into the wall with no positive connection between the end of the joist and the brickwork. To prevent the risk of injury to occupants and passers-by, loose brickwork and joists with insufficient bearing would have to be propped and the structure would be likely to continue to deteriorate unless remedial action were taken. If a home owner who is aware of a problem fails to take reasonable action and a passer-by is injured, this may give rise to serious consequences and may leave the home owner legally liable.

Is serviceability affected?

As a general rule, foundation movement is likely to impair serviceability before any part of the building becomes unsafe. Thus doors and windows are likely to stick before any cracking has appeared. Although you can normally ease and adjust these so that they remain functional, continuing or seasonal foundation movement will make this increasingly onerous. Once the damage reaches **Category 2**, you are likely to have to fill cracks to prevent wind and rain penetrating. Whether or not this is acceptable as a long-term solution depends more on how rapidly the cracks open or close than on their width. At large movements (those corresponding typically to **Category 3** damage) there is a risk of more serious defects, such as fractured service pipes and slipping roof tiles.

Will I be able to sell the house?

Your main worry may be about the effect that obvious cracks can have on the value of your home, even where there is no other significant effect. In principle, it is unreasonable for the value of an older property to be reduced because of the appearance of small cracks during periods of unusually dry weather. For a Victorian property founded at shallow depth on a shrinkable soil, for example, a limited amount of cracking at such times is quite likely. A competent surveyor should be able to distinguish damage of this kind from more serious problems and advise clients accordingly.

Similar damage in a newer property, say one built since 1970, is more unusual because the foundations should be sufficiently deep to protect the structure from the effects of dry weather. In such cases monitoring may be needed to establish whether the damage is simply seasonal or a symptom of a more serious problem. While it is likely to be very difficult for you to sell the house during the monitoring period, many insurers will for a house that is being investigated transfer cover to a new owner; this removes the danger of the new owner finding it impossible to get cover and should, in theory at least, limit the effect of the damage on the sale price.

Aesthetics

Damage usually begins to concern the average home owner long before it affects serviceability or stability. In such cases there will be a wide range of opinions on what is and what is not acceptable, depending on individual perceptions and expectations. To some owners the regular reappearance of a 1 mm crack may be totally unacceptable, whereas others may be unconcerned by a crack of 3 mm or 4 mm; indeed, an external crack of this size may go unnoticed for many years. If you own a Victorian house, it is clearly less reasonable to expect it to be totally free from cracks in walls than it would be for a modern house.

What should I do?

Most houses suffer some cracking during their lives and this can have a wide range of causes apart from foundation movement. In the vast majority of cases, these cracks require no more than filling with a suitable *mastic* or grout. The key question in your mind must therefore be: when do I need to do something more? Below we give some general guidance on what to do, based on the BRE categories of damage in Table III.

Category 0 or 1: Damage is unlikely to be more than cosmetic, and it is often difficult to identify the cause; persistent cracks can nevertheless be irritating and a building surveyor should be able to recommend steps to reduce the risk of the cracks recurring or ways to mask them using suitable wall finishes. As a general rule, however, filling the cracks will be the only action necessary.

Category 2: Cracks that appear after a long period of dry weather should not cause concern. They will normally close during the subsequent winter and you can then fill them; they should not reappear until another period of unusually dry weather. Brick arch lintels and other vulnerable details may need checking to make sure that no bricks have become dislodged. Pruning large trees or shrubs close to the affected area will reduce the chance of the cracks reappearing. Again, suitable wall finishes can mask persistent cracks.

But if cracks appear suddenly for no obvious reason or show no sign of closing during winter, you should seek professional advice (see **Making a claim**) and if appropriate, advise your insurer.

Any damage that meets more than two of the criteria listed in Table II (p.36) is likely to have been caused by foundation movement and you should advise your insurers accordingly. Reporting the damage does not oblige you to make a claim, and you may tell your insurers that you simply intend to repair the cracks yourself. But it does give them the opportunity to inspect the damage and take any action they feel necessary to prevent the damage worsening (and therefore being more expensive to repair), or to pursue a claim against a third party such as the owner of a tree on a neighbouring property, or confirm that no other action is needed. Many insurers will now pay for the cost of investigating damage directly, if the damage is consistent with a subsidence problem which has occurred within the duration of their policy; this means that you can obtain expert advice on whether or not the problem is serious without incurring any expense, (But see later, **Making a claim**, p.48).

Category 3: Cracks are probably due to foundation movement and you are likely to have noticed serviceability problems. Underpinning will not necessarily be required to stabilise the foundations. For example, where damage has been caused by the growth of a tree, it is often possible to stabilise the foundations by removing or reducing the size of the tree as described in **Tree management**, p.32. Whatever the circumstances, specialist advice is likely to be needed.

Category 4 or 5: Damage is likely to have seriously impaired serviceability and, in extreme cases, there may be a risk of instability. Urgent action may be needed to prevent the structure from becoming dangerous and remedial measures in the form of underpinning or partial rebuilding are likely to be required to reinstate the property to its original condition. You should seek expert advice promptly. Where there is a risk of any injury to occupants or passers-by, your local authority is empowered under the Buildings Act 1984 (or the Dangerous Structures section of the London Building Act [Amendment Act] 1939, if in inner London) to compel you to make it safe or to deal with the situation themselves.

Chapter 5. Making a claim

Most household buildings insurance policies cover damage caused by clay shrinkage and swelling, provided this has occurred since the start of the policy. However, the onus of proving when the damage occurred and that it was caused by an insured risk rests with you as policy holder. The burden is not heavy and can often be discharged by providing a report from an engineer or surveyor demonstrating that the damage is consistent with subsidence. If everything points to a subsidence problem the Insurance Ombudsman generally feels that the burden is on the insurer to disprove subsidence. However, where there is doubt about cause of damage the policy does not cover the cost of 'proving' the claim. You are initially responsible for any professional fees, (except where insurers have assumed direct responsibility for those costs) although in practice you can expect to be reimbursed for any reasonably incurred fees provided that the claim is successful and the professional fees were incurred with the insurers' prior agreement. Some policies have a specific limit against professional fees — either 10% or 20% of the sum insured. You will, in any case, be responsible for costs up to the excess for heave and subsidence claims, normally either £500 or £1000. In this regard, some insurers will deal with professional fees directly without deducting the policy excess; in other words, the excess is applied only against the cost of reinstatement.

Professional advice

A suitably-qualified professional will be needed to advise you on the probable cause of the damage, the scope of the investigation that is needed, and remedial options. Although the advice given should be the same whether or not the damage is covered by an insurance policy, it is clearly important that you are aware of the type of work for which the insurer is likely to pay. The vast majority of disputes arising from claims for subsidence and heave are caused by the home owner's expectations being raised unreasonably at an early stage of the investigation. It has become increasingly common over the past 20 years for engineers and surveyors to specify underpinning for houses that have suffered even minor cracks. Insurers, on the other hand, are beginning to heed the advice of technical bodies such as BRE[13] that, in most instances, underpinning is not essential to maintain the building in a satisfactory structural condition. They are increasingly likely to sanction underpinning only if it can be shown to be the most cost-effective way of dealing with the damage. **Criteria for underpinning, p.81**, deals with this subject in more detail.

In many cases a single professional will see the claim through from start to finish, including the specification and direction of repairs and remedial work. He or she therefore has a critical role to play in ensuring that the necessary work is carried out efficiently and without committing you to unnecessary expense. Unfortunately the skills required to do this work do not fall neatly within any one of the professional disciplines within the building industry. In practice, the investigation is likely to be conducted by a member of one of the following four professions.

Loss adjuster

Many insurance claims involve a firm of loss adjusters. Their primary role is to ensure that the claim is settled equitably and in accordance with the terms of the insurance policy.

Although loss adjusters are paid by insurance companies, they fulfil an impartial role. Most loss adjusting firms belong to the Chartered Institute of Loss Adjusters (CILA) and qualified individuals will have the letters ACILA or FCILA (associate or fellow of the CILA) after their names. They may also be associates or fellows of the Chartered Insurance Institute (ACII, FCII) or the Chartered Institute of Arbitrators (ACIArb, FCIArb). Individual loss adjusters are often, also, qualified surveyors or engineers.

Building surveyor

A building surveyor is someone who is expert in the repair and maintenance of houses and other buildings; building societies insist that a property is inspected by a surveyor prior to approving a mortgage. A qualified surveyor will be an associate or fellow of the Royal Institution of Chartered Surveyors (ARICS or FRICS). However, this qualification covers a range of disciplines, such as quantity surveying and valuation, in addition to building surveying.

Not all surveyors are specialists, and many home surveys may be conducted by a 'general practitioner'.

Structural engineer

A structural engineer specialises in the design and use of masonry, steel, concrete and timber as construction materials. Although some structural engineers specialise in the design of large structures, those who have their own company or work in small partnerships will often concentrate on the repair, alteration and refurbishment of houses and other low-rise buildings. Many structural engineers have considerable experience of repairing houses damaged by subsidence and heave, including the design of underpinning schemes. A qualified structural engineer will be a member or fellow of the Institution of Structural Engineers (MIStructE or FIStructE).

Civil engineer

A civil engineer specialises in building and construction generally. There is therefore considerable overlap between this profession and structural engineering and many engineers are members or fellows of both the Institution of Structural Engineers and the Institution of Civil Engineers (MICE, FICE). One of the main differences between the two is that civil engineering embraces construction in the broader sense of the word including, for example, tunnels, excavations, embankments, foundations and roads, rather than just buildings. A civil engineer will therefore have at least a working knowledge of how soil behaves and some will be specialist geotechnical engineers. This is obviously relevant to the design of foundations and to the remedy of damage caused by shrinkable clay.

As well as being MICE or FICE, geotechnical engineers are likely to be members of the British Geotechnical Society, which carries no chartered status. Geotechnical engineers tend to work for civil and structural engineering consultancies or for site investigation contractors, although some work for firms specialising in geotechnical engineering or as sole practitioners.

There is a growing tendency among insurers to recommend the use of either a chartered structural or chartered civil engineer; this is on the basis that these engineers are able to consider the full range of options for stabilising the foundations and will not specify underpinning unnecessarily.

Some engineering consultancies have both geotechnical and structural expertise in house. This is an attractive feature since geotechnical specialists are ideally suited to analysing the properties and behaviour of the soil. Unfortunately information on a firm's experience is difficult to obtain, except by personal recommendation. Often the only help you will receive is to be given the names of professionals, usually civil and structural engineers, who practise locally, by the insurers or the loss adjuster.

The professional institutions, which are listed in Appendix B, will issue a list of their members who practise in a particular locality, but will not generally give information on members' areas of expertise. Some do, however, produce directories which summarise the experience of individual members and the range of services offered by the companies for which they work. Two directories which are particularly relevant are the *Geotechnical Directory*, produced every two years by the British Geotechnical Society; and the *Ground Engineering Year Book*, produced by the Institution of Civil Engineers. These directories can be bought directly from the institutions or through some bookshops; you may also be able to order a loan copy through a local library.

More than one professional may be involved in the investigation. For example, the damage may have been discovered by an architect or surveyor employed by the home owner to do other work. In such cases, the architect or surveyor will often want to be responsible for the investigation and will enlist the services of specialists as and when they are required. Although this can prove satisfactory, it is inevitably more expensive because of the duplication of fee and is thus unpopular with insurers.

In an effort to improve the way claims are investigated and settled, most insurers now take responsibility for appointing the investigator, either directly or through a loss adjuster. In this way, insurers and loss adjusters can build up a collection of professionals they know to be suitably qualified and experienced. The main advantage for the home owner is that you are relieved of the responsibility for the costs of the investigation. The disadvantage is that you no longer have your own professional adviser. However, this should be far outweighed by the fact that you can be confident the investigator is experienced in carrying out the necessary investigations.

Most insurance companies will not object if you insist on appointing your own investigator provided that the proposed person is suitably qualified. The costs of any reasonably incurred professional fees should then be accepted as part of the settlement of the claim, if agreed beforehand. In some instances, you may want to appoint an investigator before making a claim, to advise on whether or not the damage is caused by foundation movement, for example. In general, however, you should ask your insurer's advice before you do this.

Where insurers are prepared to fund an engineering input from the outset at their own expense and you want to bring in a second opinion, then the cost will not automatically be met by the insurers. Much depends on the reason why you want a second opinion and again you are only entitled to recover professional costs if these have been approved by the insurer beforehand. The policy wording often refers to 'incurred with our agreement'.

What to expect

The first step in processing a claim for heave or subsidence damage is normally for the insurer to appoint a loss adjuster, and for the loss adjuster to visit the damaged property and establish that the damage is consistent with a subsidence problem. If it is, the loss adjuster will either advise you to appoint an investigator or will appoint one on behalf of the insurer. Where the damage is relatively minor the loss adjuster, if suitably qualified, may perform the complete investigation, appointing an engineer only at a later stage to specify and administer the remedial work. The loss adjuster will advise the insurer what the claim is likely to cost and whether or not the sum insured is adequate.

The loss adjuster will also wish to consider the age of the damage and whether it has occurred while the present insurance policy has been in operation. He or she will also consider the answers given on the proposal form when the insurance was originally taken out by the home owner and whether or not this was a true reflection of events at that time. If the insurance policy has specific claim conditions, the loss adjuster will consider whether these conditions have been complied with and if not, what has been the consequence of the failure to comply.

Following the first visit, you will normally deal with the loss adjuster rather than the insurer, unless there is a problem (see **What**

if things go wrong?, Chapter 12). If you employ your own investigator, you will need to copy any reports, invoices and correspondence to the loss adjuster; conversely, if the investigator is engaged by the insurer, or the adjuster has acted as an engineer in the first instance, you will need to request copies of technical reports from the loss adjuster.

In many cases the investigation is a protracted process, which may go on for 18 months to two years. Ironically, it is the cases of serious damage that tend to be resolved most quickly. This is largely because, where the damage is relatively minor, monitoring is often needed to establish whether work is needed to stabilise the foundations. In exceptional circumstances, where the damage is very unsightly, insurers may pay for temporary repairs (e.g. filling of internal cracks and painting of the affected section of wall) to restore the property to an acceptable standard while the investigation is carried out. Generally, however, repairs will only be sanctioned once it has been established that there is a reasonable expectation that any future foundation movements will not cause further significant damage.

Once the investigations have been completed the investigator will put forward his or her proposals for repairs and any remedial measures that are needed to stabilise the foundations. The repairs will normally be aimed at reinstating the property to its former, undamaged condition and will therefore include, for example, the redecoration of any rooms that have suffered damage.

The remedial measures proposed by the investigator should have a reasonable expectation of preventing further significant damage; they will not guarantee that further damage will be prevented. In many cases, there may be more expensive solutions (e.g. extensive underpinning) with a lower risk of further damage occurring, and there may be cheaper schemes with a higher risk. As far as is practicable, the investigator should point out the range of options available and their relative costs and expectations. The final choice will then rest with whoever is funding the work. An insurer may opt for a remedial scheme with a high probability of preventing further damage in order to avoid the expense and inconvenience of further claims. More commonly, however, the insurer will be prepared to accept that there is some risk of the damage recurring. In such circumstances, it would be inappropriate for the insurer to refuse to renew existing cover or to apply a second excess in the event that you need to make a further valid claim for damage attributable to the same cause (see **What if things go wrong?**, p.99).

Once the scope of any remedial works has been agreed, the work is usually carried out by a building contractor. The procedures involved at this stage are described in **Having the work done (Chapter 11)**.

Chapter 6. The investigation

The investigation should answer four questions.

- Have the foundations moved?
- Is the cause clay shrinkage or heave, some other cause, or a combination of factors?
- Is the movement continuing, or is there potential for further movement?
- If so, what can be done to reduce further damage?

To answer these questions it is essential that the investigator identifies the cause of the damage as precisely as possible. To do this, he or she will collect information on soil type, the type and depth of the foundations, the pattern of movement of the building, the presence of trees, the presence of water, and the nature of the surrounding topography.

Deciding whether or not further movement and damage is likely may require a period of monitoring during which movement of the building is measured at regular intervals.

Because cost will be a constraint, the essence of a good investigation is to identify at an early stage whether there is a serious problem, so that sensible decisions can be made about the need for, and cost-effectiveness of, further investigations.

This chapter describes the components of a typical investigation. These are: an initial visual inspection; a distortion survey; a desk study; trial pits and boreholes; and a drain survey. They should enable the investigator to decide on the cause of the movement. The next chapter looks at monitoring.

Initial inspection

The first stage is normally a visual inspection to establish the seriousness of the damage, its probable cause and the need for further investigations. Usually this is carried out after a formal claim has been lodged with insurers, although in some instances it may be performed to help the home owner decide whether he or she has a legitimate claim. The essential elements of the inspection are the same in either case.

The initial survey of the damaged property should include a sketch showing the position, width and taper of all internal and external cracks,[12] as shown in Figure 18. There are two reasons for doing this: first, it provides an objective record against which future damage can be compared and second, it helps to identify the pattern of movement (that is, the way in which any foundation movements are affecting the building as a whole). This is important, because it may highlight damage that is being caused, for example, by rotation of a wall, which may therefore be well removed from the area that is subsiding or heaving. In other words, the location of the damage does not necessarily correspond to the location of the foundation movement. The severity of the damage should be evaluated according to the classification given in Table III (p.39).

It is important to establish the history of the damage and you can play an important role by providing information about when cracks first appeared and whether they have remained at their original size, opened progressively or opened in summer and

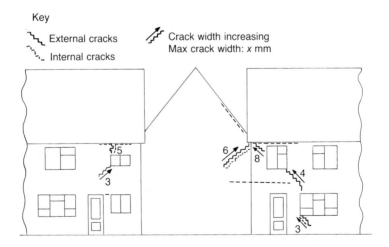

Fig. 18. Recommended method of recording damage. (BRE. Crown Copyright)

Fig. 19. Using a portable water level to measure out-of-level of brick course. (BRE. Crown Copyright)

closed in winter. In addition, the investigator will look for signs of previous repairs and will examine the surfaces of cracks: recent cracks in brickwork tend to have a clean appearance, whereas dirt will have accumulated in older cracks.

The size, position and species of all trees close enough to have an effect on the foundations will also be noted. Ideally, the initial inspection should also include a distortion survey.

Distortion survey

The purpose of the distortion survey is to measure the amount of any foundation movement that has occurred. Although it may not be needed in every investigation, it offers two important benefits: firstly it can provide incontrovertible evidence that the foundations have moved or, conversely, show that it is unlikely that they have; secondly it defines the size and extent of any movements.

The simplest way of quantifying the movement is to measure the change in level along a course of bricks. Where the external walls are rendered so that brick courses cannot be seen, an indication of the amount of movement that has occurred can be gained by measuring any deviation of the walls from the vertical. Either of these measurements can be made using relatively simple equipment. For example, verticality can be measured using a plumb-line and ruler[14] and changes in level can be measured using a *portable water level*, which is essentially a plastic bottle connected to a graduated cylinder by a length of flexible tubing. Using a portable water level, it takes about one or two hours for one person to check brick course levels around a detached house. Figures 19 and 20 illustrate the application of these techniques.

Other types of distortion survey may be appropriate;[15] for example, where it is suspected that damage is caused by swelling soil, it may also be useful to measure changes in level across the floor slab.

Preliminary report

The investigator should now be in a position to state whether he or she believes the damage is the result of foundation problems and the probable cause. A preliminary report should set this out; it will

Fig. 20. Using a plumb-line and ruler to measure verticality (BRE. Crown Copyright)

also summarise the scope of further investigations needed and the range of possible remedial options. The report should make clear what damage the investigator believes has occurred since the start of the insurance cover.

Provided the inspection has been carried out by a qualified professional, this report should in most cases determine whether or not the insurer accepts that there is a legitimate claim. If the claim is considered valid, the initial inspection and any subsequent investigations can be considered part of the remedial process and should be paid for by the insurer.

Having decided the foundations have moved, further action may be needed to establish the potential for further movement, and to provide information for the remedial work. Such action can include desk studies, the excavation of trial pits and boreholes, and a drain survey.

Desk study

A desk study comprises a search and review of existing data and such reference material can often provide useful clues to the cause of foundation movement. As a bare minimum, the local geology should be checked by reference to the 1:50,000 drift or drift and solid geological map, or, where available, the six inch to one mile or 1:10,000 series. In some cases[6] it may be a good idea to check old maps, records and aerial photographs for the locations of pits, streams, trees, and other features that may have had an influence on the foundations. It may also be possible to find copies of the plans for the original building work and any extensions, which can then be used to check the intended depth of foundations, any precautions taken against heave when the house was built, floor slab design e.g. reinforced or unreinforced, wall floor connections, and other construction details. This course of action is cheaper and less disruptive than having to excavate *trial pits*, although there is always the snag that the actual construction may have deviated from that shown in the plans; foundation details, in particular, are likely to be modified during construction to meet the ground conditions encountered and consequently foundation depths are often shown on the plans as 'to be agreed on site'.

(a) Light cable percussion rig

(b) Hand-held drive-in sampler

(c) Restricted access rig

Fig. 21. Three types of sampling equipment : (b) and (c) courtesy of Abbey Underpinning and Foundations; (a) BRE. Crown Copyright

Trial pits

Small excavations or trial pits are often needed to confirm the depth and condition of the foundations and to establish the nature of the underlying soil. Shallow samples taken from a trial pit can be used for tests of soil properties, such as plastic and liquid limits, which can then be used to establish the soil's shrinkage potential (see **What are shrinkable clays**, p.13). Often it is possible to obtain deeper samples using a *hand auger* and these can then be used to measure the variation of moisture content or suction with depth, and hence to obtain an indication of the depth of any desiccation.

Once it has been confirmed that some foundation movement has occurred, it is generally a good idea to excavate at least one trial pit early in the investigation. In straightforward cases, this may be the only soil testing that is required.

Boreholes

Mechanical boring is expensive and should be used sparingly. It is likely to be justified, for example: where removal of a large tree is being considered and it is important to determine accurately the depth and degree of desiccation; where a house is suspected of being on a filled pit and it is important to confirm the geology to considerable depth; and where a pile-based underpinning scheme is being considered and it is important to profile properties such as the strength of the soil for design purposes.

When investigating shrinkable clays, samples of undesiccated clay should, wherever possible, be obtained from a control borehole for comparison. In certain circumstances, for instance where the clay is heterogeneous or no control borehole is possible, the only way of reliably identifying desiccation is to measure the *suctions* in the soil. (see **How much desiccation** p.16).

Various types of drill rig are used for obtaining borehole samples, some of which are illustrated in Figure 21. A light cable percussion rig is often used to obtain 'undisturbed' 100 mm diameter samples which are needed to determine the strength of the soil accurately. Where this is not important, a powered auger or hand held drive-in sampler is likely to be cheaper. These are also smaller and can be used in restricted spaces.

Drain survey

As explained previously (see **Drainage**, p.33), water leaking into the ground near foundations is generally undesirable and the investigator will often want to check the condition of the drains to help eliminate the possibility of leakage having contributed to any foundation movement. Manholes situated within your property are used to gain access to the drains. Leakage can be detected by performing a series of simple pressure tests on sections of drain that have been isolated using inflatable packers or expandable plugs.

These tests are carried out by flooding the manholes and measuring the time taken for the water to escape.

Alternatively, the drains can be examined visually using a small close circuit television (CCTV) camera. The camera survey is likely to be more expensive than pressure tests but has the advantage of being able to identify the nature of the defect and to pinpoint its location.

What caused the movement?

By now the investigator should be able to say, in most cases, what has caused the foundations to move, and to make recommendations for remedial work. Though shrinkage and swelling of clay soils are the most common causes of foundation movement, they are not the only ones. Factors the investigator will take into account in reaching a conclusion follow.

Soil type Only clay soils expand and contract as their moisture content changes. If the soil is mainly granular (silt, sand, gravel,

Fig. 22. Undersailing of brickwork below damp-proof course caused by swelling clay. (BRE. Crown Copyright)

or chalk, for example) volume change can immediately be ruled out.

Foundations Susceptibility of a building to ground movements depends chiefly on the depth of its foundations. If no trees are present, clay shrinkage is unlikely to damage a building founded at a depth of 1 m or more. Foundation depth may provide clues about the site conditions at the time of construction. If foundations in a clay are deeper than the standard 0.9 m, it is likely that the builder was anticipating large ground movements, perhaps as a result of trees cleared from the site prior to construction. This can be checked by looking for deep roots in boreholes or by studying aerial photographs of the site taken before construction. The use of piles, pads or a raft foundation suggests that additional movement due to difficult ground conditions was anticipated.

Pattern of movement Looking at the pattern of cracking, aided by measurements from a distortion survey, will usually establish which way the ground has moved — up, down, or sideways - to cause the damage. Some types of damage are characteristic of specific causes — for example *undersailing* of brickwork below damp proof course, as shown in Figure 22, usually denotes swelling soil.

Trees As already mentioned, by influencing desiccation nearby trees play a crucial role in causing ground movements.

Water Ground water running freely into a trial pit or borehole in a clay soil indicates the presence of layers of coarse-grained soil. This can affect the permeability of the soil dramatically and will allow desiccation levels to change more quickly than in a pure clay. If free water is present for any length of time, desiccation is highly unlikely.

Surroundings Local topography, such as sloping ground, and impermeable surfaces of concrete or asphalt, can affect the moisture content of the soil (see Chapter 2).

In some cases, there may be several contributory factors. For example, a property may be more susceptible to damage as a result of clay shrinkage because it settled differentially during the first few years following construction. Similarly, damage unrelated to clay shrinkage may be exacerbated as the result of seasonal volume changes in the soil. It is therefore essential to gauge whether or not the scale of the damage and distortion is consistent with the assumed cause. For instance, if clay shrinkage is thought to be responsible the question the investigator must address is: are the pattern and size of the foundation movements compatible with the depth of the foundations, the properties of the soil, and the size, type and position of any nearby trees? If not, then other processes must be involved and further investigation may be needed to identify them. These could include further desk studies, a more detailed soil investigation, or monitoring.

Chapter 7. Monitoring

What is monitoring?

Once it has been confirmed that foundation movement has caused the cracks, insurers and loss adjusters are increasingly calling for measurements to be made over a period of time to show whether or not the movement is continuing; this is called *monitoring*. Unfortunately, because it inevitably delays settling the claim, and the reasons for the delay are often not appreciated, monitoring has frequently caused acrimony between policy holders and their insurers.

In fact, monitoring is a very powerful tool and in many cases is the fairest and most objective way of establishing whether or not a damaged property needs to be underpinned. However, monitoring is not a substitute for a thorough investigation and it should only be used in cases where the results are likely to be of benefit. These cases fall into three broad categories, each with a particular approach to monitoring.

Monitoring to establish cause of damage

Where an investigation, as described previously, has proved inconclusive, monitoring can be a very cost-effective diagnostic tool. For example, where a house is founded on shrinkable clay, but there is no correlation between the distortion and the position of nearby trees, monitoring can distinguish movement due to seasonal clay shrinkage, which tends to be cyclic, from that due to processes such as settlement, landslip and erosion, which tend to develop in one direction.

Monitoring to measure rate of movement

Where the cause of the damage is self-evident, monitoring can be used to establish whether the damage is continuing to worsen, and if so whether the rate of movement is slowing down. This can be a very useful technique where the damage has been caused by a process that has a limited duration, such as heave following removal of a tree. Similarly, where the damage has occurred during a period of abnormal weather, this type of monitoring can help establish whether the foundation movements in a 'typical' year are likely to be tolerable.

Monitoring to check success of remedial action

Where action has been taken to remove the cause of the damage, such as cutting down or pruning nearby trees, monitoring can gauge the effectiveness of the remedy. The investigator can then decide whether or not further work such as underpinning is needed.

Monitoring should only be considered if it is unlikely that the condition of the property will deteriorate significantly during the period of observation; it is therefore important that the measurement techniques are accurate enough to detect changes before they have any noticeable effect on the building. Where the initial observations indicate that the damage is worsening rapidly, it will be necessary to consider the need for immediate remedial work without waiting for the end of the monitoring period.

Monitoring techniques

A monitoring programme may concentrate on one of two aspects: measuring the damage to see if it is getting worse (or better); or measuring movement of the building to see which parts of the foundations are moving and at what rate. In addition, it is sometimes desirable to measure the movements in the ground directly; for example, where there appears to be some horizontal movement, measurements made in the ground are likely to be more conclusive than measurements made on the structure.

The various techniques that can be applied to monitoring damaged buildings are described briefly in the following sections.

Crack width monitoring

The most common and simplest way of monitoring subsidence damage is to measure changes in the width of existing cracks. This can be done in several ways.[16]

A GUIDE TO SUBSIDENCE AND HEAVE OF BUILDINGS ON CLAY 67

- *Steel rule* Provided sufficient care is taken, crack widths can be measured to the nearest 0.5 mm using a steel rule. However, because the readings tend to be subjective and it is difficult to ensure that the crack is measured at the same point each time, this method is normally used only for recording the state of damage during the initial inspection.

- *Magnifier and graticule* Internal cracks in plaster or other smooth finishes can be monitored by measuring the offset between two pencil marks using a magnifying glass fitted with a graticule, as shown in Figure 23. With care, movements can be measured to an accuracy of 0.1 mm.

- *Glass tell-tales* Cementing glass strips across cracks, as shown in Figure 24, used to be a popular method of detecting progressive movement. However, such tell-tales give little indication of how much movement is taking place and are easily vandalised. Consequently, the use of this technique should be avoided.

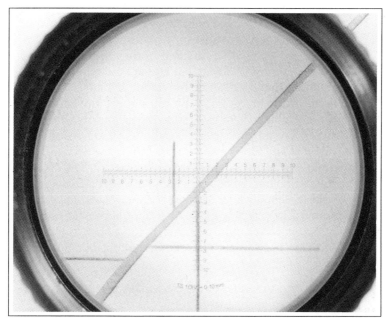

Fig. 23. Measuring crack width using a magnifying glass and graticule. (BRE. Crown Copyright)

Fig. 24. Crack monitoring using a glass tell-tale. (BRE. Crown Copyright)

- *Plastic tell-tales* The most popular system is shown in Figure 25. It consists of two overlapping plates screwed to the wall, one marked with a cursor, the other with a scale graduated in millimetres. The two plates are mounted on opposite sides of the crack so that the cursor is initially in line with the centre of the scale; any subsequent movement of the crack can then be measured to the nearest millimetre on the scale. The advantage of this system is that a reading can be taken at any time by anyone, including the occupiers, without any additional measuring equipment. The disadvantages are that the tell-tales are relatively obtrusive, vulnerable to vandalism or accidental damage, and have only a limited accuracy. In general, they should be used only in conjunction with a more accurate method, such as the 'Demec' or 'brass screw' techniques described below.

- *Demec points* Two small, dimpled stainless steel discs are fixed on opposite sides of the crack. The distance between them can then be measured very accurately with a separate, hand-held instrument called a 'Demec' gauge. This gauge was developed for measuring very small movements — as little as 0.02 mm — in concrete and masonry in laboratory tests. A disadvantage is that it is so accurate that it will also detect movement due to changes in temperature and moisture in the brickwork, which can confuse the picture. Demec points have the advantage of being very unobtrusive, but they are only suitable for use on a flat surface and cannot be used to measure cracks at corners. The main disadvantage, however, is the limited range of the gauge: the maximum extension that can be measured is 2.4 mm and the maximum contraction 1.6 mm. Where significant movement is occurring, it will be necessary to install replacement discs at intervals to allow readings to continue.

- *Brass screws* The technique recommended by BRE[16] is to fix small brass screws into the wall either side of the crack and to measure the distance between them using a caliper. This system has the advantages of being simple, robust, relatively unobtrusive and, by using the calipers in different modes, capable of measuring cracks in corners and other

Fig. 25. Crack monitoring using a plastic tell-tale. Courtesy of Avongard

HAS YOUR HOUSE GOT CRACKS?

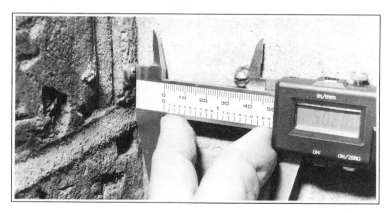

Fig. 26. Crack measurement using brass screws (BRE. Crown Copyright)

awkward positions, as shown in Figure 26. If three screws are arranged in a right-angle triangle, both horizontal and vertical movements can be measured. With a digital caliper, an accuracy of better than 0.1 mm overall should be easily achievable.

Where screws might be obtrusive, particularly internally, Demec studs can be used instead and the measurements made by locating the tips of the caliper jaws in the central dimples. Plastic tell-tales, also fitted with Demec studs, are now available to enable more accurate measurements to be made using calipers.

Monitoring levels

However accurately crack widths are measured, these are the symptoms and not the cause. Hence the results can be ambiguous; a crack may form for one reason and progress for another. Once a crack is formed, stresses in the masonry from other causes, which in themselves were not enough to cause a crack, will redistribute themselves and may widen the crack, even though no further foundation movement has taken place. Hence, in most cases, it is better to measure the vertical movements of the foundations as well as crack widths.

This can be done using a precision optical level to record the movements of monitoring points fixed to the building. It is essential to use a precision instrument for these measurements to achieve an overall accuracy of ±0.5 mm. For most applications, small screws or masonry nails can be used as monitoring points; these are normally sufficiently unobtrusive to avoid acts of vandalism.

Wherever possible, levels should be measured relative to a fixed reference point or datum. For most domestic applications, a stormwater drain or similar deep feature is sufficiently stable for this purpose. Where there are no deep drains or where absolute accuracy is imperative, a deep datum can be specially installed at a suitable depth.[17]

Monitoring lateral movements

Lateral movements in shrinkable clays are rarely measured because of the costs involved. The usual technique is to install a special plastic tube in a borehole about 15 m deep. Movement of the tube from the vertical can then be measured by lowering an inclinometer down the tube. The technique is highly specialised and should only be performed by a firm with the necessary expertise.

Observation period

Monitoring normally needs to continue for a period of at least a year to distinguish seasonal movements from long-term subsidence or *recovery*.

However, in most cases where damage has been caused by clay shrinkage or swelling, the benefits of longer periods of monitoring are likely to be limited. The observation period should not, therefore, exceed say 18 months without very good reason. Ideally readings should be taken every month, but in practice every six weeks for crack monitoring or every two months for level monitoring is usually adequate.

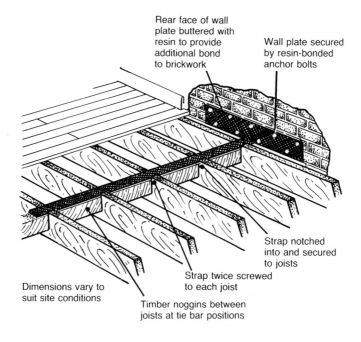

Fig. 27(a). Repair using tie bar. Courtesy of Falcon Repair Services Ltd

Chapter 8. Preventing further damage

If it is confirmed that the foundations have moved and that further movement is likely, a decision has to be made on how to prevent further damage. One solution is to underpin the foundations.

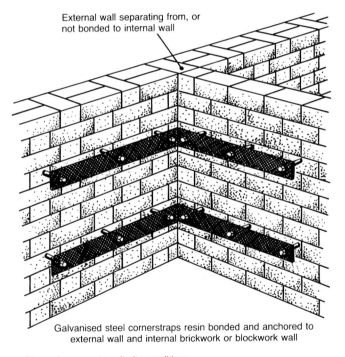

External wall separating from, or not bonded to internal wall

Galvanised steel cornerstraps resin bonded and anchored to external wall and internal brickwork or blockwork wall

Dimensions vary to suit site conditions

Fig. 27(b). Repair using corner straps. Courtesy of Falcon Repair Services Ltd

(a) Before   (b) After

Fig. 28. Resin bonding of brickwork. Courtesy of Abbey Underpinning and Foundations

However, underpinning is a relatively drastic and often expensive solution and other options should be explored first. These include strengthening the superstructure, reducing the influence of nearby trees and stabilising the soil.

Repairing or strengthening the superstructure

Where it can be shown that the cause of the damage is a process which is now largely over, such as heave following removal of a tree, or one that is likely to occur only rarely, such as clay shrinkage during exceptionally dry weather, it is generally possible to prevent further damage by repairing or strengthening the superstructure. Techniques include use of tie bars and straps (see Figure 27), resin bonding of brickwork (see Figure 28), brickwork stitching (see Figure 29), and mortar bed reinforcement (see Figure 30). Details of traditional repair techniques can be found in any good text book on structural repair, such as Melville and Gordon's *The repair and maintenance of houses*.[18]

A relatively new technique, known as *corseting*, consists of casting a reinforced concrete beam around the perimeter of the building, usually at or below ground level. The beam is connected to the brickwork by means of vertical steel reinforcing bars or 'dowels'

A GUIDE TO SUBSIDENCE AND HEAVE OF BUILDINGS ON CLAY 75

and the beam is subsequently tensioned by a torque wrench or hydraulic jack. The corset stiffens the building at foundation level, and helps it bridge local areas of subsidence.

Reducing the influence of trees

Where heave has been caused by the removal of trees, there is nothing that can be done to prevent the swelling process running its full course. However, where shrinkage has been exacerbated by trees, one of the following techniques may reduce their influence and provide a very cost-effective remedy.

Tree removal

Removing the tree altogether will have the greatest and most immediate effect on the levels of desiccation in the soil. As explained earlier, this should be safe provided the tree is no older than any part of the house, since the consequent heave can at worst only return the foundations to their original level. In most cases there is no advantage to a staged reduction in the size of the tree and the tree should be completely removed at the earliest opportunity.

The time taken for the soil to recover depends largely on the permeability of the soil. In heavy clays, such as London Clay, it may take tens of years for the ground to reach equilibrium, even though most of the heave occurs during the first few years; in one well-

(a) **Before**

(b) **After**

Fig. 29 Brickwork stitching. Courtesy of Falcon Repair Services Ltd

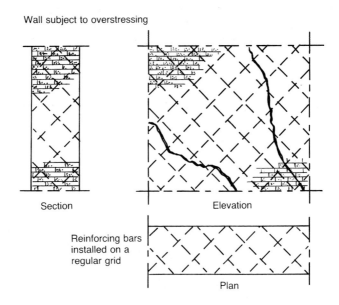

Fig. 29(c). Brickwork stitching Courtesy of Falcon Repair Services Ltd

documented case,[19] where large elm trees were removed from a London Clay site prior to construction of some cottages, movements were still measurable 25 years later, as shown in Figure 31. In more permeable soils, full recovery may be achieved in a few years. The length of time which recovery is likely to take may be a factor in deciding whether or not removing the tree is an acceptable solution.

Where the tree is older than the house, or there are more recent extensions to the house, it is not advisable to remove the tree altogether because of the danger of inducing damaging heave. In such cases, your investigator should calculate the *heave potential* in the soil adjacent to the foundations before deciding whether or not the tree can be removed.

Tree pruning

Where it is unsafe to remove the tree altogether and the cracking is relatively minor, some form of pruning, such as *crown thinning*, *crown reduction* or *pollarding* should be considered. Pollarding, in which most of the branches are removed and the height of the main trunk is reduced, is often mistakenly specified, because most published advice links the height of the tree to the likelihood of damage. In fact the leaf area is the important factor. Crown thinning or crown reduction, in which some branches are removed or shortened, is therefore generally preferable to pollarding. The pruning should be done in such a way as to minimise the future growth of

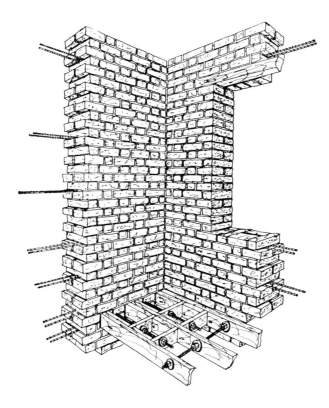

Fig. 30. Mortar bed reinforcement. Courtesy of RME Ltd

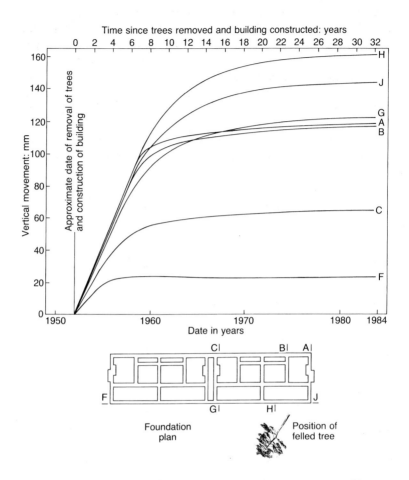

Fig. 31. Observed heave of some cottages built on a London. Clay site cleared of trees prior to construction. (BRE. Crown Copyright)

the tree, without leaving it vulnerable to disease (as pollarding often does) while maintaining its shape. This should be done only by a reputable tree surgeon or qualified contractor working under the instructions of an *arboriculturist*.

Root pruning

You may find there is opposition to the removal or reduction of an offending tree; for example, it may belong to a neighbour or the local authority, or have a Tree Preservation Order on it. In such cases there are other techniques that can be used from within your own property. One option is root pruning, which is usually performed by excavating a trench between the tree and the damaged property deep enough to cut most of the roots. The trench should not be so close to the tree that it jeopardises its stability.[3] In time, the tree will grow new roots to replace those that are cut; however, in the short term there will be some recovery as the degree of desiccation in the soil under the foundations reduces. Where the damage has only appeared in a period of dry weather, a return to a normal weather pattern may prevent further damage occurring.

Permission from the local authority is required before pruning the roots of a tree with a preservation order on it.

Root barriers

Root barriers are a variant of root pruning. However, instead of simply filling the trench with soil after cutting the roots, the trench is either filled with concrete or lined with an impermeable layer to form a 'permanent' barrier to the roots. Whether the barrier will be truly permanent is questionable, because the roots may be able to grow round or under the trench. However, the barrier should at least increase the time it takes for the roots to grow back.

Soil stabilisation

For some causes of subsidence it is possible to prevent further movement by improving the stability of the underlying soil; for example, cavities in the soil caused by erosion can be filled by injecting cement-based grout under pressure. The stabilisation of clay soils that are moving as a result of changes in moisture content is less straightforward. In theory, the characteristics of the clay and its tendency to change volume can be significantly altered by adding certain chemicals. Shrinkage potential in particular can be reduced by using lime which replaces sodium ions in the clay minerals with calcium ions. This technique is effective in the laboratory and can be used to treat clay fill, but the extremely low permeability of most shrinkable clays makes the technique extremely difficult to use in the ground. This severely limits the usefulness of chemical additives, although some proprietary treatments are available.

Another technique that is sometimes used in an attempt to reduce seasonal shrinkage and swelling is artificially to increase the supply of water. Gravel-filled trenches may help reduce desiccation levels locally provided they are maintained full of water. Unfortunately, unless large quantities are used, any water added during the summer is likely to evaporate before it infiltrates the soil enough to have any lasting benefit. It is also unwise to site drains too close to foundations, as there would be a danger of softening the soil and causing worse settlement.

Remedial underpinning

Where existing foundations are believed to be inadequate, they can be stabilised by *underpinning*, which means either providing new foundations or, more usually, extending the existing foundations downwards to reach stiffer or more stable ground. Chapter 9 sets out how to decide whether or not underpinning is the correct solution; Chapter 10 explains how underpinning is done.

Chapter 9. Is underpinning the right solution?

The use of underpinning as a remedy for subsidence and heave damage has increased dramatically since insurance cover for these risks was introduced. However, in 1991 a BRE report[13] concluded that most underpinning is not technically justified and this may be at least partly responsible for the growing reluctance among insurers to accept underpinning solutions.

Where the remedial measures described in Chapter 8 can be used, they will generally be more cost-effective than underpinning and will provide an equally valid solution. But where there is severe damage, or nothing else is thought likely to be effective, underpinning will often be the correct solution and should be specified without delay.

Criteria for underpinning

Although there is no generally accepted method for deciding when underpinning is justified, your investigator will consider the following factors.

Is structural stability threatened?

Where the damage is so severe that there is doubt about the ability of the building to continue to carry the loads applied to it (in other words, where there is risk of walls, floors, or the roof becoming unstable), urgent action will be needed to prevent it collapsing. Although external shoring or internal propping can be used as a temporary solution, a permanent solution will almost certainly require underpinning or partial rebuilding on deeper foundations; in extreme cases, the most cost-effective solution may be demolition and rebuilding of the whole structure. This is unlikely to apply unless the damage is Category 5 (Table III, p.39); it is rare for damage due to movement in shrinkable clay to be as severe as this.

Is movement continuing?

Where the structural stability of the building is not threatened, the prognosis for further damage becomes of prime importance. Engineers, loss adjusters and insurers will want to know whether the movement is 'progressive' — in other words, is there evidence, such as cracks widening, of increasing damage as a result of continuing foundation movement?

Where there is, the foundations must be stabilised. However, this does not necessarily imply underpinning. As mentioned in Chapter 8, damage caused by a large tree may be effectively dealt with by pruning it and repairing the cracks.

On the other hand, where heave has been caused by removing a tree, there is nothing that can be done to prevent the ground from swelling and if further substantial movement is expected, underpinning will almost certainly be the best option.

Is the movement excessive?

Changes in crack widths can give a false impression of what is happening to the foundations. A more fundamental approach to deciding whether underpinning is needed is to measure foundation movements directly (see **Monitoring levels**, p.71) and make a judgement about whether or not they are 'excessive'. Since the technique for making the measurements is straightforward and inexpensive, the only drawback with this approach is the lack of definitive guidance on how much movement is excessive. A rational approach would be to compare the observed movements of the damaged house with measurements made on similar undamaged properties over the same period. A database of typical foundation movements would therefore provide a useful benchmark against which monitoring data could be judged and this may evolve as the use of level monitoring increases. In the meantime, the interpretation of monitoring data will continue to depend on engineering judgement.

What is the cost?

Because foundation movements associated with shrinking or swelling clay are unlikely to threaten the structural stability of the building, the arguments for or against underpinning are primarily related to cost and removal of the blight of repairs at intervals in the future. Some people argue that the cost of underpinning to be

A GUIDE TO SUBSIDENCE AND HEAVE OF BUILDINGS ON CLAY

Damage classification	Appropriate action (Relevance of underpinning in italics)
0 to 1	Remedial measures are generally unnecessary as cracks can be repaired as part of routine maintenance. Where cracks recur during periods of dry weather, consider pruning nearby trees and shrubs. Monitoring is needed to confirm that damage is caused by foundation movement. *Underpinning unlikely to be cost-effective except in very rare circumstances, for example where there is recurrent damage to expensive wall finishes.*
2	Cracks which appear at end of summer and close during subsequent winter can be repaired in spring and steps taken to reduce the risk of damage recurring, such as pruning nearby trees and shrubs. Where cracks are not seasonal, having taken steps to minimise the movement, monitoring should be used to establish extent, magnitude and rate of foundation movement. *Underpinning is unlikely to be cost-effective, unless foundation movement is progressive or excessive and there is either a likelihood of recurrent damage which will be expensive to repair; or the potential for further movement (e.g. as a result of heave) will create excessive damage (say Category 4)*.*
3	Having taken steps to mitigate the cause of the movement, monitoring should be used to establish extent, magnitude and rate of movement; brick arches and other susceptible features may need propping to prevent deterioration. *Underpinning is likely to be cost-effective, where movement is progressive or excessive and alternatives such as tree removal are impracticable.*
4	Unless there is a risk of instability, monitoring should be used to establish extent, magnitude and rate of movement. Wherever practicable, steps to remove the cause of the movement should be taken prior to monitoring. *Underpinning is needed to prevent instability where movement is progressive or excessive, unless the cause of the damage is obvious and can be easily removed; for example, if caused by a large tree and there are no impediments to its removal. This may be preferable to underpinning.*
5	Temporary support (e.g. external shoring and/or internal propping) is probably needed to prevent collapse. Monitoring may be needed to give warning of instability, but is unlikely to aid selection of appropriate remedy. *Underpinning or rebuilding on deeper foundations needed to reinstate affected areas†; work should be implemented rapidly to prevent unnecessary deterioration of the structure.*

Table IV. Appropriate action for various levels of damage
* as defined in Table III
† in some circumstances lifting or jacking the structure back to level may provide an economic alternative to rebuilding.

justified must be less than the cost of periodically repairing the cracks.

In practice, applying the principle is less straightforward than this because the chance of the damage recurring depends on unpredictable factors such as the weather, and because it is difficult to quantify inconvenience and distress to the home owner in financial terms. Nevertheless, using engineering judgement, comparisons on economic lines can be made, and these should be given considerable weight when stability or safety is not an issue.

Is underpinning needed?

As a general rule, many insurers and loss adjusters will not recommend underpinning unless the damage exceeds Category 2 in Table III, but will readily approve it where damage is Category 4 or 5. For Category 3, the decision will depend very much on the likely effectiveness of options such as tree removal.

Table IV gives a suggested framework for deciding whether a property should be underpinned. It should be stressed, however, that this table has been compiled by the authors and has yet to be ratified by any insurer. In practice, every insurance company is developing its own criteria for deciding whether or not underpinning is justified. In many cases these criteria are not definitive and the decision is likely to be coloured by a variety of other considerations, including the history of the property and whether there have been previous claims, and your own circumstances and personal preferences.

Chapter 10. Which type of underpinning?

There are essentially four underpinning methods that can be used: mass concrete, pier-and-beam, pile-and-beam and mini-piling.[13,20] The choice of method is governed primarily by two factors: the ground conditions and the required foundation depth.

Mass concrete

Underpinning using mass concrete is often referred to as 'traditional' underpinning because the principle has been in use for centuries. In the past, when labour costs were low and before ready mixed concrete became widely available, traditional underpinning was constructed in brickwork; nowadays, mass concrete is invariably employed.

In principle, mass concrete underpinning is a method of deepening existing strip or pad foundations so that they reach a stratum with adequate bearing properties. The underpinning is carried out in a series of bays or areas as shown in Figure 32. The width of each bay is determined by the ability of the walls to span the gap created; for most houses with competent brick or stone-work, this is likely to be in the range 1.0 to 1.4 m. Where there is any doubt about the wall's ability to span the bay, the wall should be *needled* and the load transferred to temporary supports bearing on the ground.

Groups of bays with the same number (see Figure 32) are excavated at the same time, so that no more than 20% to 25% of the wall is left unsupported at once. The mass concrete is cast into the bay to leave a 75 mm to 150 mm gap between the concrete and the underside of the existing foundation. Once the concrete has had a minimum of 24 hours to harden, this gap is 'pinned up' by ramming in a moist concrete with a maximum aggregate size of 10 mm and just enough water to enable the mixture to remain in a ball when squeezed in the hand. This mix will have little tendency to shrink;

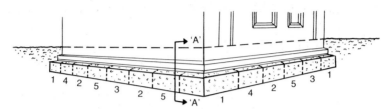

The bays are numbered to indicate a typical sequence of excavation, concreting and 'pinning up'

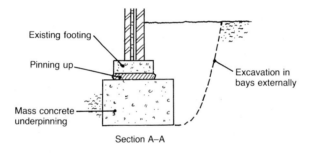

Section A–A

Fig. 32. Mass concrete underpinning

ramming it in tightly creates intimate contact between the old and new foundations and prevents settlement.

After the first group of bays is completed, work can begin on the next group. It is normal practice to allow at least 24 hours between pinning up and the excavation of an adjacent bay. The procedure is repeated until the entire building has been underpinned.

A variant of traditional mass concrete underpinning is staggered or 'hit-and-miss' underpinning. Instead of forming a continuous strip beneath an existing footing, the underpinning remains as discrete bays or piers. The span between piers is determined by the strength of the existing footings. This method would not normally be used where foundation loads are relatively high or where the footings are shallow, insubstantial or cracked as a result of ground movement. The use of hit-and-miss underpinning is therefore somewhat limited, but it can be an economic solution in favourable circumstances and is especially suitable where existing foundations are constructed of reinforced concrete.

The simplicity of mass concrete underpinning means that it can be installed by a relatively non-specialist workforce and is therefore available from sources such as local builders. Nevertheless, careful specification and direction of the site work by a qualified engineer are considered essential in order to obtain a satisfactory result. The cost advantage of mass concrete underpinning declines rapidly as the required depth increases, because of the increasing costs of materials and transporting spoil away, and the high cost of hand excavation. Although mass concreting to depths of 4 m or more is possible, as a general rule it is unlikely to be the cheapest option where the depth required exceeds 2 m; this limits its use in cases where subsidence or heave is caused by a large tree. Mass concrete may also be unsuitable where there is a surface layer of loose or waterlogged ground which would hinder hand excavation.

Pier-and-beam

Pier-and-beam or *base-and-beam* underpinning was introduced shortly after the Second World War by the underpinning specialist Pynford and is now widely used. It is useful where the depth of the foundation needed is too great for traditional underpinning to be economic. Instead, isolated mass concrete piers are dug to the required depth at intervals round the building, with a reinforced concrete ground beam at or above footing level spanning between the piers and supporting the walls (see Figure 33).

Pier-and-beam underpinning is feasible in most ground conditions. However, it tends to be economical only at depths shallower than about 4 m. Excavations can be carried out where there is loose or waterlogged soil by using shields (e.g. trench sheeting), but this adds considerably to the cost. Pier-and-beam underpinning is particularly suitable for use in shrinkable clay where further volume changes are anticipated. The piers can be excavated to depths at which the effects of shrinkage and heave are minimal and, provided the sides of the piers are protected, the building can be isolated from the effects of further volume changes in the soil. The ground beam serves to strengthen the building early in the underpinning process.

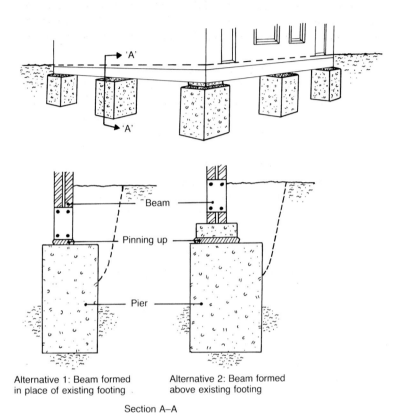

Alternative 1: Beam formed in place of existing footing

Alternative 2: Beam formed above existing footing

Section A–A

Fig. 33. Pier-and-beam underpinning

Pile-and-beam and piled-raft

Pile-and-beam systems are similar in concept to pier-and-beam underpinning, but have advantages where there is no suitable bearing stratum available within a depth economical for hand excavation of piers (about 4 m to 5 m) or where it is necessary for the underpinning to pass through loose or water-bearing strata.

Ground beams are normally constructed in the same way as for pier-and-beam underpinning, except that the beams are extended at corners and intersections to form caps for attaching to the pile heads. Intermediate supports are formed by pairs of piles with

A GUIDE TO SUBSIDENCE AND HEAVE OF BUILDINGS ON CLAY

'needle capping beams' or, where internal access is restricted, using 'cantilever pile caps' as shown in Figure 34. It is preferable to install the piles after the construction of the beams, especially where there is a risk of pile installation causing further damage to the unstrengthened building. This makes detailing and construction of pile caps difficult.

Piles vary in diameter from 150 mm to 400 mm for low-rise buildings, although smaller diameters may be used on lightly-loaded structures. Because of access limitations and the need to minimise vibration which could affect vulnerable structures nearby, the piles are usually constructed by augering a hole in the ground in which concrete is then cast to form the pile. However, smaller diameter piles may be driven.

Where external access is restricted, the existing floor slab can be removed and the piles installed inside the house; they are then connected using a reinforced concrete raft which is keyed to the

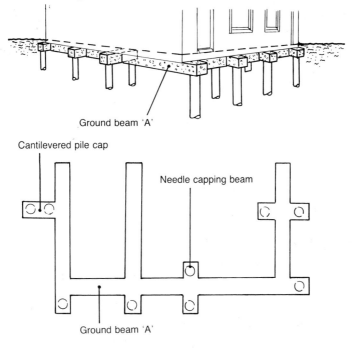

Fig. 34. Pile-and-beam underpinning

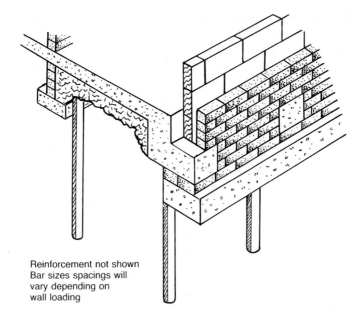

Reinforcement not shown
Bar sizes spacings will
vary depending on
wall loading

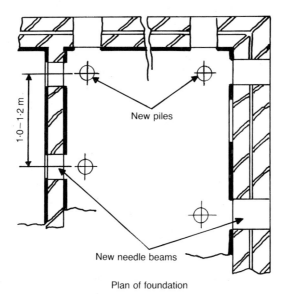

Plan of foundation

Fig. 35. Piled raft underpinning. Courtesy of Roger Bullivant Ltd

external walls below ground level by removing small sections of brickwork, as shown in Figure 35. This is called a *piled-raft system*.

Piled-raft systems are attractive to the underpinning designer where the floor slab has to be replaced in any case to avoid damage as a result of clay heave.

One disadvantage of pile based systems is that the relatively slender piles (150 mm to 400 mm diameter) used in domestic scale underpinning provide less resistance under lateral loading than more robust piers. In addition, deeper and more sophisticated ground investigation data are needed for the proper design of the piles.

Mini-piling

In this method of underpinning, there is no ground beam. Loads are transferred directly from the structure to the piles, either by needles or cantilevered pile caps, or by placing the piles directly through the existing footing. In low-rise buildings the pile supports would normally be at 1 m centres or less. Because of the short spans between piles, pile loads are low and small diameter *mini-piles* are used. Typically, diameters range from 65 mm to 150 mm. They are formed either by driving steel or plastic casings into the ground and then filling them with cement grout, or alternatively they can be bored and cast in situ. Which technique is adopted depends largely on the ground conditions. Guidance on the design, supervision and approval of remedial works and new foundations for low-rise buildings based on mini-piles is given in BRE Digest 313.[21]

Mini-pile systems are distinct from pile-and-beam and piled-raft underpinning, because they rely on the strength of the existing foundations to transmit the wall loadings. Consequently, mini-piling tends to be cheaper than other pile-based systems and can even compete with mass concrete underpinning on cost. However, because of the small diameter of the piles, they are unsuitable for applications where high lateral loads are envisaged, as is often the case in shrinkable clay. Moreover, because mini-piles can only be used where the strength and integrity of the existing foundations can be assured, they are unsuitable for older properties or properties in a bad structural condition.

Mini-piles are particularly suited to underpinning buildings on uniform, shallow thicknesses of fill and natural soils not susceptible

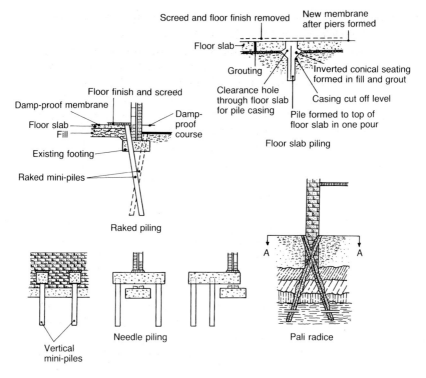

Fig. 36. Four different types of mini-piling

to shrinkage and heave. They are also one of the few techniques that can be used to stabilise floor slabs (see Figure 36).

Partial underpinning

Where only part of the building has been affected by ground movement, it is generally unnecessary to underpin the whole structure. An underpinning scheme that does not include all load-bearing walls is called partial underpinning. In some cases, the underpinning will be restricted to one side of the property; in others, the internal walls may be left in their original condition. Underpinning of a semi-detached or terraced house, where it may not be possible to extend the remedial scheme to the neighbouring properties, is also in effect partial underpinning.

Partially underpinning a building on shrinkable clay needs special care. As shown in Figure 5, page 21 foundations at a depth of 1 m may move by up to 13 mm in a dry year. This is not normally noticeable because the whole house will move as one unit. But if part of the building is underpinned, it is likely to move less than the non-underpinned part. These differential foundation movements can cause further damage where the underpinned and non-underpinned sections meet.

One way of avoiding such damage is to extend the underpinning under the unaffected part of the building and to reduce the depth in steps to avoid creating hard spots. This is most easily achieved with mass concrete underpinning, where the depth of each bay can be easily varied; the usual practice is to decrease the depth of underpinning in 0.3 m steps until it merges with the original foundations.

This approach can also be applied to pier-and-beam and pile-and-beam systems, by extending the ground beams beyond the last pier or pile under the unaffected part of the house.

Alternatively, the likelihood of cracking can be reduced by ensuring that the depth of the underpinning is not over-specified, so that the underpinned section of the building continues to move more or less in sympathy with the rest of the structure. This may increase the risk of existing cracks re-opening slightly, but may provide a more cost-effective solution than, for example, having to underpin the whole structure.

Chapter 11. Having the work done

The procedure for having the work carried out is the same whether the damage is being repaired under an insurance claim or you are paying for it yourself. In either case, a suitably qualified professional should be employed to prepare a specification and to direct the work. Although this may sometimes be an architect or surveyor, if your house is being underpinned an engineer will be required.

Specification
The professional adviser will draw up a schedule (or list) of work to be done, which will fall into three broad categories:

- groundworks: underpinning, rebuilding foundation brickwork, grouting, replacing floor slabs, etc.

- structural repairs: removing and rebuilding sections of brickwork, resin injection or stitching of cracks, installing tie rods and steel straps, repointing mortar joints, etc.

- making good: plastering, repapering or retiling walls, adjusting doors and windows, painting.

Where the work is being done under an insurance claim, you will have to agree the scope of the work with the loss adjuster. This often involves an element of compromise, particularly over decoration. For example, to repair cracks, a bathroom or kitchen may have to be retiled; the insurance will only cover the cost of providing and fitting similar tiles. If you want to use a better quality tile you should expect to pay the extra in the cost of the materials.

The contract
Remedial or repair work will normally be carried out under a standard contract, such as the Minor Works Agreements issued by

the Joint Contracts Tribunal (JCT) or the Institution of Civil Engineers. If no formal agreement is signed, you should at least exchange letters with the contractor confirming the terms under which the work is to be performed. The contract is between you and the contractor, and does not involve the insurer. Once work begins, though employed by you, the professional adviser is expected to act impartially in settling any disputes which arise between you and the contractor.

As well as defining the scope of the work, the contract will specify the extent of the contractor's liability. In some circumstances, you may need to make arrangements to extend your insurance cover to include accidental damage to adjacent properties as a result of the building work. The consequences of failing to arrange special insurance can be severe, and professional advice should be sought. Often the loss adjuster will be able to offer some advice regarding the arrangement of adequate insurance cover during the course of the work, otherwise you may need to approach an insurance broker. Additional costs which arise through obtaining extra insurance are normally recovered as part of the claim.

The contract will also define how the contractor will gain access to the site, and in some cases it may be necessary to get permission from neighbours for some of the work to be carried out from their land.

Where a semi-detached or terraced house is being underpinned, the owners of the adjoining house or houses should be informed, so that they can take any action to safeguard their interests. Where a party wall is to be repaired or underpinned, it is advisable to prepare a formal Schedule of Condition for the neighbouring property to avoid any disputes regarding damage resulting from the work. For certain London boroughs this action is mandatory under the provisions of the London Building Act (1939); this act requires the adjoining owner's permission to allow any reinforced foundations to be placed under the party wall.

A contractor for the work is usually selected by competitive tender. The main contract is usually let to a general builder, with provision for separate sub-contracts for specialist work such as the underpinning itself, or decoration. Alternatively, many underpinning firms now offer a wide range of services including structural repairs and making-good, and could therefore perform the complete contract in-house. Typically three or four contractors would be chosen to tender for each contract or sub-contract.

It is clearly important to select a competent and reliable contractor and the professional adviser will ensure that tenders are sent only to firms capable of completing the necessary work satisfactorily. If you have personal knowledge of a reputable builder you can, of course, ask for his or her name to be included in the list of tenderers. Once the tenders have been received, the professional adviser will recommend which contractor should be given the work. This will normally be the one who has submitted the lowest bid, unless any of the contractors have attached unreasonable conditions to their tender returns or have failed to include all the necessary items specified in the Schedule of Work.

If the work is being funded as part of an insurance claim, the professional adviser will pass his recommendations both to you and to the loss adjuster acting for your insurers. Although the ultimate choice of contractor is yours, insurers will normally only pay for the lowest acceptable tender bid. Therefore, if you choose one of the more expensive contractors, you should expect to pay the extra cost yourself.

Will I have to move?

Where extensive underpinning or structural repairs are being carried out, for safety or convenience you may have to move out while the work is being done. It is not uncommon for a period of three to six months to be needed to complete the work. The cost of alternative accommodation, or loss of rent in the case of rented accommodation, is covered by most buildings insurance policies which stipulate limits to the amount of money available. The standard of alternative accommodation provided is normally based on your minimum requirements, rather than being on a like-for-like basis. Accommodation and loss of rent costs are shared between buildings and contents insurers.

Payment

Because the contract is between you as home owner and the builder, you are responsible for making the necessary payments. However, where the work is being done as part of an insurance claim, you will normally receive payments in stages from your insurer. This will allow you to pay the interim invoices submitted by the builder; these should be certified by the professional adviser to say that the work they cover has been completed satisfactorily.

Many contractors insist that you sign a payment mandate, which allows the insurers to pay the contractor directly. Payments will only be made once they have been authorised by the professional adviser. Often insurances are arranged through a bank or building society with whom a home owner has a mortgage. It is usually a condition of the mortgage that payment is made by insurers in the joint names of the home owner and the building society, so that the society will need to countersign the cheque before it is passed on to the home owner or contractor.

When the work is virtually complete, you will normally be asked to sign an 'acceptance form', which confirms to the insurers that the claim has been resolved to your satisfaction and allows the final payment to be made by insurers.

Supervision

The professional adviser will visit the site periodically to inspect the work, record progress and sort out any technical problems. The frequency of the visits will depend primarily on the complexity of the operations that are being undertaken. For example the professional adviser might carry out daily inspections of underpinning work, but only visit the site once a week while decoration is being carried out. You should, however, bear in mind that overzealous 'supervision' of contractors is likely to interfere with their work and can therefore be counter-productive. You should be particularly careful not to issue any verbal instructions which could be construed as extending or modifying the agreed Schedule of Work; if you do, you may find yourself footing the bill for the extra work. To be part of the contract, any amendments to the Schedule of Work must be issued formally through the professional adviser.

The professional adviser can not provide day to day supervision of the work. This is normally provided by the contractor's foreman or clerk of works. For small jobs, such as decoration, there may be no formal supervision and in such circumstances it is important that you let the professional adviser know if you are unhappy with the way the work is being carried out.

Building Regulations

Any underpinning or major rebuilding work requires approval under the Building Regulations (1991),[22] or in the case of buildings in Scotland, the Building Standards (Scotland) Regulations (1991).[23] To obtain this approval, plans for the work have to be submitted to the local authority, who will then make periodic inspections of the work on site. The responsibility for obtaining Building Regulations approval may rest with the professional adviser or the contractor, and should be explicit in the contract. There are different ways in which Building Regulations approval can be obtained and your professional adviser should be familiar with them.

Warranties

As a general principle, it is impractical and uneconomic to specify remedial solutions that will absolutely guarantee that no further damage will occur. In most cases, the scheme specified by the professional adviser will have a reasonable expectation of providing a satisfactory solution considering the relative cost and likely effectiveness of the other options. The professional adviser will not, therefore, give any form of guarantee or warranty. However, in common law the professional adviser does owe his or her client a duty to exercise reasonable skill and care in executing the work.

Many contractors will issue a 10 or 20 year guarantee for the underpinning work. However, with a few exceptions, these guarantees are not backed up by insurance and will only remain valid as long as the underpinning firm remains in existence. Many guarantees also have various conditions which attach — typically excluding damage which might occur to underpinning as a result of continued growth of trees.

Chapter 12. What if things go wrong?

Problems can arise at three stages: during the investigation; while the work is being done; and at any time after the completion of the work, should the damage return. What should you do if this happens?

During the investigation
Where damage is being investigated as part of an insurance claim, you will normally deal with a loss adjuster appointed by the insurers. If you feel that the claim is taking an unnecessarily long time to deal with, or that the service is poor, you should initially address your complaint to the principal of the firm of loss adjusters or engineers, as appropriate. If you do not obtain satisfaction then the matter should be reported to the insurers. Where insurance is arranged through a building society, bank or insurance broker it may well be that these intermediaries will also be able to offer some assistance. If these actions fail to resolve the matter, you can appeal to the Insurance Ombudsman Bureau (IOB), whose address is given in Appendix B. The IOB will then give a decision on whether or not the insurers are fulfilling their obligations under the terms of the policy. This service is provided at no charge to you. However, membership of the IOB is not compulsory, though most leading insurers do belong. You will therefore need to check that your insurer is a member. Before referring a claim to the IOB, you must give the insurance company a final chance to resolve the matter by writing to, and receiving a reply from, the company's Chief Executive at head office. The IOB can deal with most disputes concerning subsidence and heave claims but remember that if your case is solely about delay, a reference to the Bureau may further hold up your claim as the insurer must send all its papers to the IOB for its investigation.

The judgement issued by the IOB is binding on the insurers but not on you. Referring the matter to the IOB does not, therefore, prevent you later pursuing the matter through the courts. However, unless you qualify for Legal Aid, the cost of legal action is generally prohibitive and should only be considered as a last resort. If you do decide to take legal action, contact a solicitor with experience of handling subsidence or heave claims. The cost of taking such advice will be at your expense and an action through the courts may take a considerable time, perhaps years.

Insurers who are not members of the IOB will usually agree to some form of arbitration. However, such an arrangement is normally binding on both parties.

During the contract

Any problems that arise during the execution of the works are normally dealt with by the professional adviser. Where he or she is unable to resolve a dispute, the contract usually has provision for the appointment of an arbitrator whose decision is binding on both you and the contractor. In exceptional circumstances, it may be necessary to replace a contractor who fails to fulfil his or her obligations.

After completion

If the remedial work fails to prevent the damage returning, insurers may deal with the damage as a new claim subject to a separate excess or a continuation of the old claim, depending on the circumstances. If the ongoing movement is due to inadequate design or workmanship in the original underpinning, then the new damage may not give rise to a valid claim. In such circumstances the home owner may need to take up the matter directly with the builder, professional adviser or possibly consult a solicitor.

It is rare for an insurer not to continue cover for subsidence and heave following the settlement of a claim. If cover is refused you should establish why this should be the case, as clearly it implies some lack of confidence in the work as carried out. Whereas an insurance company has the right to cancel or alter the contract of insurance, provided adequate notice is given, you should not be disadvantaged as a result of having made a claim. Consequently, if, by refusing to continue the cover, the insurer leaves you with a property that is uninsurable, you should consider reporting the matter to the IOB. The IOB cannot compel the insurer to offer cover, but could instruct the company to pay for further work to make the property an acceptable risk or award you compensation for the loss of value to your house caused by being unable to insure it.

References

1. **National House Building Council.** *Standards.* NHBC, Amersham, 1991.

2. **British Standards Institution.** BS 8004 *Foundations.* BSI, London, 1986.

3. **British Standards Institution.** BS 5837 *Trees in relation to construction.* BSI, London, 1992.

4. **British Standards Institution.** BS 5930 *Code of practice of site investigation.* BSI, London, 1981. (The proposed revision of this Standard is being prepared by BSI subcommittee B/526/1).

5. **British Standards Institution.** BS 1377 *Methods of testing soils for civil engineering purposes.* BSI, London, 1991.

6. Site investigation for low-rise building: desk studies. *BRE Digest* 318, 1987.

7. A method of determining the state of desiccation in clay soils. BRE Information Paper IP4/93, 1993.

8. The influence of trees on house foundations in clay soils. *BRE Digest* 298, 1985.

9. **British Standards Institution.** CP 101 *Foundations and substructures of non-industrial buildings of not more than four storeys.* BSI, London, 1949.

10. **National House-Building Council.** *Practice Note 3 — Building near trees.* NHBC, Amersham, 1969.

11. Trees near the house. *Gardening Which,* 1989, Nov., pp 359–361.

12. Assessment of damage in low-rise buildings. *BRE Digest* 251, 1990.

13. **R. Hunt, R. H. Dyer** and **R. Driscoll.** *Foundation movement and remedial underpinning in low-rise buildings.* BRE Report BR184, 1991.

14. Simple measuring and monitoring of movement in low-rise buildings: part 2 — settlement, heave and out-of-plumb. *BRE Digest* 344, 1989.

15. **P. Robson**. *Structural appraisal of traditional buildings*. Gower Technical, London, 1990.

16. Simple measuring and monitoring of movement in low-rise buildings: part 1 - cracks. *BRE Digest* 343, 1989.

17. Monitoring building and ground movement by precise levelling. *BRE Digest* 386, 1993.

18. **I. A. Melville** and **I. A. Gordon**. *The repair and maintenance of houses*. Estates Gazette Ltd., London, 1988.

19. **J. E. Cheney**. 25 years' heave of a building constructed on clay, after tree removal. *Ground Engineering*, 1988, vol. **21**, no. 5, pp 13–27.

20. **S. Thorburn** and **G. S. Littlejohn**. *Underpinning and Retention*. 2nd edition. Blackie Academic & Professional, London, 1993.

21. Mini-piling of low-rise buildings. *BRE Digest* 313, 1986.

22. Building Regulations. HMSO, London, 1991.

23. Building Standards (Scotland) Regulations. HMSO, London, 1991.

A. Glossary of technical terms used in the text

(Page numbers refer to explanations or principal usage in the text)

Arboriculturist: a tree specialist, preferably a member of the Arboriculturists Association, and who should carry appropriate liability insurance. (p.79)

Base-and-beam: a method of underpinning based on cast in-situ ground beams supported by squat concrete columns; also referred to as 'pier-and-beam' (p.87).

Beam-and-block floor: a method of constructing a *suspended floor* slab consisting of precast concrete beams and lightweight concrete blocks (see Figure 10, p.28.)

Bearing capacity: the maximum foundation load that can be applied to a soil (p.111).

Bond: the arrangement of bricks, blocks or stones within a masonry wall to a set pattern to achieve a combination of adequate strength and attractive appearance (p.30).

Corbelling: stepping of brickwork to increase the width of a wall in order to support a load (p.24).

Corseting: a method of reinforcing low-rise buildings based on installing post-tensioned concrete beams at foundation level (p.74).

Crown reduction: a method of pruning trees based on reducing the size of the canopy by shortening the length of branches (p.77).

Crown thinning: a method of pruning trees based on reducing the size of the canopy by removing selected branches (p.77).

Damp-proof course (dpc): a waterproof layer installed near the base of a masonry wall to prevent upward movement of moisture (p.40).

Desiccation: any significant reduction in soil moisture content, caused by evaporation or extraction of moisture by trees, shrubs, etc. (p.13).

Differential settlement (or differential foundation movement): a measure of the distortion in a wall based on the vertical displacement of one point with respect to another (p.5).

Footing: a shallow concrete foundation placed under a wall or column to spread the load over a larger area of ground (p.24).

Ground beam: a reinforced concrete beam used to support a wall and to transfer the wall loading to piles or pads (p.27).

Hand auger: a boring tool used to excavate holes of between 50 mm and 250 mm diameter in soil (p.60).

Hardcore: coarse inert granular material commonly used to fill hollows and to provide a suitable base on which to cast a concrete floor slab (p.28)

Heave: upward ground movement and the corresponding movement of affected foundations (p.5).

Heave potential: a quantitative measure of the capacity of a desiccated soil to generate upward movement in existing or proposed foundations (p.76).

Headers: bricks laid across a wall so that they are end on to the outside face (p.30).

Hogging: deflection of a wall or beam resulting in the ends being lower than the middle: opposite of *sagging* (p.38).

Illite: one of the three common clay minerals (p.13).

Joist: a beam, often timber, used to support floorboards or ceilings (p.44).

Kaolinite: one of the three common clay minerals (p.13).

Landslip: movement of soil down a slope (p.9).

Lime mortar: a bonding agent for masonry consisting of a mixture of sand and lime (p.30).

Lintel: a beam, usually of timber, concrete or steel or an arch of brick or stone, placed above a door or window opening to support the weight of the wall above (p.44).

Liquid limit: a measure of the minimum moisture content at which a clay looses its 'plastic' properties and begins to flow (p.14) (cf plastic limit).

Mastic: flexible sealant used to fill gaps where movement is anticipated, for example a window frame and surrounding brickwork (p.46).

Mini-pile: type of piling commonly used for *underpinning*, using driven or cast-in-place piles with a diameter of between 65 mm and 150 mm (p.92).

Moisture content: A measure of the amount of moisture contained in a sample of soil calculated from measurements made before and after drying at a temperature of 105°C; expressed in gravimetric (as a percentage of residual weight after drying) terms. (p.5)

Monitoring: periodic measurements of a damaged building to establish whether foundation movement is continuing or damage worsening (p.65).

Montmorillonite: one of the three common clay minerals (p.13).

Needle: a small beam installed under or through a wall as a support (p. 85,).

Overconsolidated: description of clay that has previously existed under far higher confining stresses than at present (p.14).

Peat: soft, compressible, dark brown or nearly black soil derived from vegetable matter (p.15).

Permeability: a measure of the rate at which water will flow through a soil; the more *permeable* a soil, the greater the flow of water under a given pressure gradient (p.13)

Pier: a squat concrete or masonry column constructed below ground level and used to support either concentrated point loads or reinforced concrete ground beams.(p.27)

Pier-and-beam: a method of underpinning using cast in-situ ground beams supported by squat concrete columns; also referred to as 'base-and-beam' (p. 87).

Pile: a relatively long slender foundation element used to transmit foundation loads to a deep stratum, fabricated from timber, concrete or steel; piles may be installed by driving or by casting concrete into a bored hole. (p.27)

Pile and beam: a method of underpinning based on cast in-situ ground beams supported by piles (p.87).

Piled raft: a method of underpinning based on a cast in-situ reinforced slab supported by piles (p.91).

Plastic limit: a measure of the minimum moisture content at which a clay retains its 'plastic' properties and does not break up when moulded (p.14) (cf *Liquid Limit*).

Plasticity index (or plasticity): the difference in moisture content between the Plastic limit and the liquid limit for a given sample of clay (p.16).

Pollarding: a method of tree management in which most of the branches are removed and the main trunk is shortened (p.77).

Portable water level: a device for measuring the vertical distance between two points (p.58).

Raft foundation: a type of foundation in which the entire building is supported on a reinforced concrete slab (p.28)

Recovery: increase in soil moisture content and associated increase in volume, produced by a reduction in the forces (or suctions) that cause desiccation (p.72).

Sagging: deflection of a wall or beam resulting in the middle being lower than the ends: opposite of *hogging* (p.40).

Settlement: downward movement of soil under load, especially as soil compresses under foundation loads (p.9).

Short bored piles: a type of foundation for houses or other low-rise buildings using relatively short (usually less than 10 m long) piles constructed by boring a hole and filling it with concrete (p.11).

Shrinkable clay: a clay, the volume of which changes significantly with variations in moisture content (p.13).

Shrinkage potential: a qualitative measure of a clay soil's capacity to cause damage as a result of volume change; three classifications exist: 'low', 'medium' and 'high' (p.16). cf *heave potential.*

Silt: a soil made up of particles with diameters of size intermediate between clay (less than 0.002 mm) and sand (greater than 0.06 mm) (p.13).

Sleeper wall: a low wall constructed at foundation level to support a suspended timber floor (p.29).

Stools: temporary supports usually made of steel, which are used to carry wall loads to allow sections of brickwork to be removed during underpinning operations (p.68).

Strip footing: a shallow concrete foundation cast in the bottom of a trench to provide continuous support for a wall (p.11).

Stretchers: bricks laid in the line of the wall so that the long side is visible on the face of the wall (p.30).

Stud partition: method of constructing non load-bearing walls in houses in which plasterboard is attached to a timber frame. (p.40).

Subsidence: downward ground movement and the corresponding movement of affected foundations (p.5).

Suction (or pore water suction): negative pressure within the water occupying the spaces between soil particles, which can be caused by evaporation from the surface of the soil and the extraction of moisture through the roots of vegetation and is therefore commonly associated with the process of desiccation (p.61).

Suspended floor: method of constructing a ground floor so that it is not supported on the ground (p.29).

Trench-fill foundation: type of foundation commonly used for houses and other low-rise buildings, where a narrow concrete-filled trench is used to provide continuous support for a load-bearing wall (p.11).

Trial pit: a small excavation dug to inspect foundations and/or soil conditions; the maximum depth for a hand dug trial pit is about 1.5 m, although depths of 3 m or more are easily achievable with a mechanical excavator (p. 59).

Underpinning: a technique for replacing or deepening existing foundations (p. 80).

Undersailing: the outward movement of brickwork below the *damp-proof-course* (p.63).

Void former: collapsible material used to form a space under cast-in-place concrete floor slabs in order to protect the slab from the effects of swelling soil (p.29).

B. Professional organisations

Building Research Establishment (BRE)
Advisory Service
Garston
Watford
Hertfordshire WD2 7JR
Telephone 01923 894040

British Geological Survey (BGS)
London Information Office
Geological Museum
Exhibition Road
London SW7 2DE
Telephone 0171 589 4090

British Geotechnical Society (BGS)
1 Great George Street
London SW1P 3AA
Telephone 0171 222 7722

Chartered Institute of Loss Adjusters (CILA)
376 Strand
London WC2R 0LR
Telephone 0171 240 1496

Institution of Civil Engineers (ICE)
1 Great George Street
London SW1P 3AA
Telephone 0171 222 7722

Institution of Structural Engineers (IStructE)
11 Upper Belgrave St
London SW1X 8BH
Telephone 0171 235 4535

Insurance Ombudsman Bureau (IOB)
135 Park St
London SE1 9EA
Telephone 0171 928 7600

National House-Building Council (NHBC)
Buildmark House
Chiltern Avenue
Amersham
HP6 5AP
Telephone 01494 434477

Royal Institution of Chartered Surveyors (RICS)
12 Great George St
London SW1P 3AE
Telephone 0171 222 7000

Appendix C. Foundation and superstructure design

New build — The easiest way to avoid subsidence damage is to make sure that the foundations of new buildings are adequate in the first place.

A certain amount of foundation movement is inevitable, since ground will compress under the loads applied by the foundations; equally, ground movements may occur as a result of processes that are unconnected with the applied loads, such as changes in the moisture content of shrinkable clays. It is generally both impractical and uneconomic to design foundations to be totally static throughout the life of the building. A successful foundation design will therefore ensure that the level of movement transmitted to the superstructure is acceptable and that distortions never exceed tolerable levels.

In many instances foundation design depends on the *bearing capacity* of the underlying soil. For buildings founded on firm, shrinkable clays, however, this is unlikely to be an important consideration, because such clays are strong enough to support a low rise building on a conventional strip footing of, say, 400 mm width. Rather it is the depth of the foundations that is critical for building on these soils — they should be deep enough not to be affected by changes in moisture content. The effects of evaporation and moisture extraction by vegetation reduce with depth. Where there are no trees or large shrubs, a foundation depth of 0.9 m is generally adequate. But foundations designed to this minimum requirement allow no provision for future tree planting. In practice, many home owners will, at some point, want to plant small trees such as fruit trees near the house and so it may be prudent, though costly, to provide foundations that are deeper than the recommended minimum. A depth of 1.5 m, for example, would allow an apple or pear tree to be safely planted at half its mature height (4 m to 6 m) from the foundations even on highly shrinkable soil.

Clearly it may be necessary to use even deeper foundations where there are existing trees on the site, especially if these are large broad leaf trees such as oak, willow, elm or poplar. Recommended foundation depths for different types of tree, at various distances from the foundations, and for three classifications of shrinkable soil are contained in the NHBC Standards.[1] Since the maximum recommended depth is 3.5 m, it is possible to use trench-fill foundations in all circumstances. The depth of desiccation under a large tree can be considerably greater than the recommended foundation depth; for example, in London Clay, desiccation to depths of 6 m is not unusual. Nevertheless, the degree of desiccation tends to decrease with depth and tends to be fairly constant in the deeper soil unless the tree is removed or grows substantially larger. Consequently, the depth of the foundations can be substantially less than the depth of desiccation and still provide adequate stability for the house. It follows that large trees should be left in place wherever possible. This would not apply where the tree is on, or very close to, the proposed site for the house and, in such circumstances, a piled foundation should be considered. This can be designed to give a far higher margin of safety against movement and is often no more expensive than deep trench-fill.

A fundamental disadvantage of deep trench-fill foundations in heavily desiccated soils is that, by cutting through tree roots, they inevitably upset the equilibrium in the soil even if no trees are removed; this in turn generates lateral movements in the soil, which then tend to push the foundations sideways. To reduce the effect of the lateral pressures it is necessary to protect one side of the trench (normally the inside face) with a compressible material such as low-density expanded polystyrene. However, in such circumstances, many engineers prefer to use piled foundations since piles have less of an effect on the equilibrium in the soil and may readily be taken down below the desiccated zone.

Floor slabs are also susceptible to damage as a result of clay shrinkage and swelling and, when building on shrinkable soils, it is advisable to use a suspended floor with an adequate void under it, rather than a slab bearing on the ground. The NHBC Standards[1] offer advice on the depth of void required; for example, 150 mm is recommended for soils with a high shrinkage potential.

Extensions

The foundation requirements for extensions are essentially the same as those for new houses, with one important proviso. With new houses, the distortions due to foundation movements are limited by the fact that the whole structure tends to be affected at the same time. In the case of extensions, however, any initial settlement will show itself as differential movement between the extension and the original house and, where the foundation design of the extension is different from that of the original house, the response of the two parts to dry weather may be dramatically different. It follows that problems may arise wherever the foundations for the extension are significantly deeper than those of the original house.

The foundations for any extensions will have to comply with the Building Regulations (1991),[22] or, in the case of buildings in Scotland, the Building Standards (Scotland) Regulations (1991).[23] These regulations require compliance with existing codes and guidance and, in the case of building on shrinkable clay, this is likely to be interpreted as meeting the recommendations of the NHBC Standards.[1] Consequently the minimum foundation depth for an extension founded on clay will be 0.9 m and may be substantially greater where there are large trees in the garden. Equally, if the house is more than 30 years old, it may have been built on foundations that are no more than 0.6 m deep and may in many cases be even shallower. In such cases, the most important consideration is avoiding creating a 'hard spot' in the structure, where the foundation depth changes rapidly over a short distance. There are three ways in which this can be achieved:

- underpin some of the foundations of the original house to the level required for the extension and step the underpinning up progressively away from the extension to avoid creating a 'hard spot';

- match the foundations of the extension to those of the original house and step them down progressively away from the original house in line with the NHBC recommendations; or

- build the extension entirely on deep foundations and provide a movement joint between the two structures.

None of these options will guarantee a satisfactory result in all cases and it will be necessary to consider the merits and cost of the three options in each case before deciding on the appropriate foundation design.

Superstructure details

In theory, quite a lot can be done to reduce the susceptibility of low-rise buildings, such as houses, to damage from foundation movement. For example, framed structures with brick infill could be used in place of load-bearing masonry, or reinforcement could be inserted into mortar courses near weak spots such as window openings. In practice, this approach is rarely used because it is generally more straightforward to ensure that the foundations perform adequately.

One exception to this rule is where a sudden change in foundation depth is unavoidable, either because of varying ground conditions or because an extension is being added. If it is not possible to step the foundations up to avoid creating a hard spot, then a vertical movement joint can be introduced into the structure. This is basically a gap in the brickwork filled with a flexible sealant (sometimes described as a *mastic*), typically around 10 mm wide.

Movement joints are essentially deliberate cracks. As such they help prevent the walls from cracking elsewhere and, provided they are properly constructed, should not detract from the overall appearance of the house and should provide adequate protection from rain and wind. The most common use of movement joints is in large panels of brickwork, where they are essential to accommodate movements associated with changes in temperature and the initial expansion of the bricks. The use of movement joints as a means of avoiding damage where an extension abuts an original house is becoming increasingly common.